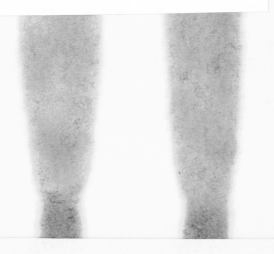

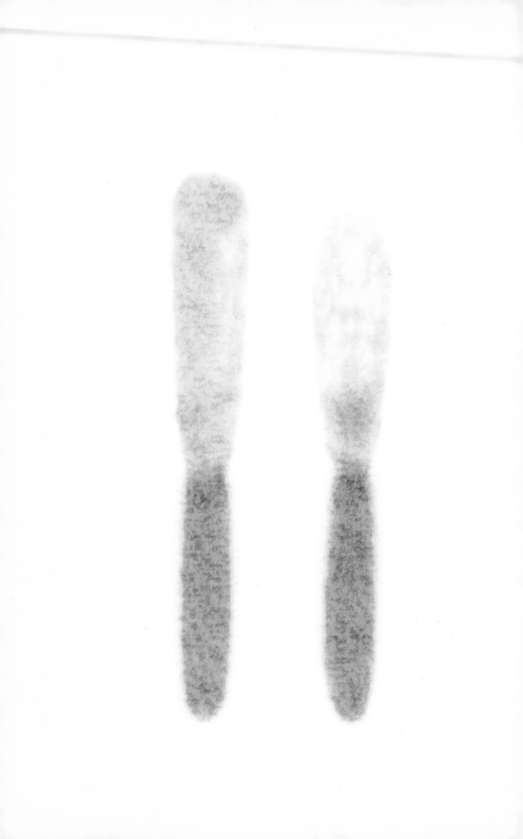

comap, inc.

870605

The UMAP Expository Monograph Series

UMAP Monographs bring the undergraduate student new mathematics and fresh applications of mathematics with no delay between the development of an idea and its implementation in the undergraduate curriculum.

The High Cost of Clean Water:
Models for Water Quality Management
Edward Beltrami, *SUNY at Stony Brook*

Spatial Models of Election Competition
Steven J. Brams, *New York University*

Modeling Tomorrow's Energy System:
Applications of Linear Programming
T. Owen Carroll, *SUNY at Stony Brook*

Elements of the Theory of Generalized Inverses for Matrices
Randall E. Cline, *University of Tennessee*

Introduction to Population Modeling
James C. Frauenthal, *SUNY at Stony Brook*

Smallpox: When Should Routine Vaccination be Discontinued?
James C. Frauenthal, *SUNY at Stony Brook*

Conditional Independence in Applied Probability
Paul E. Pfeiffer, *Rice University*

Topics in the Theory of Voting
Philip D. Straffin, Jr., *Beloit College*

Markov Decision Processes
Paul Thie, *Boston College*

Man in Competition with the Spruce Budworm:
An Application of Differential Equations
Philip M. Tuchinsky, *Ford Motor Company's Research and Engineering Center*

The UMAP Expository Monograph Series

Markov
Decision
Processes

Paul Thie
Boston College

consortium for mathematics
and its applications, inc.
271 lincoln street, suite no. 4
lexington, ma 02173

Author: Paul Thie
          Mathematics Department
          Boston College
          Chestnut Hill, MA   02167

This material was prepared with the partial support
of National Science Foundation Grants No. SED80-07731
and No. SPE-8304192. Recommendations expressed are
those of the author and do not necessarily reflect
the views of the NSF or the copyright holder.

© COMAP, Inc. 1983
ISBN 0-912843-04-7
Printed in USA

Contents

Preface

In a Markov decision process, we have, as in a Markov chain, a system that can move from one distinguished state to another. However, at each step, a decision maker effects the transition probabilities for the system by selecting, from a well-defined set of alternatives, a particular course of action. Associated wih each action is also a gain (or loss). The obvious problem facing the decision maker is to determine a suitable plan of action to follow over the duration of the operation so that the total gain is optimized.

The Markov decision process model has become a useful operations research tool. The model has applications in such diverse areas as equipment maintenance, inventory control, and investment analysis. And there exist effective solution techniques which can be programmed to handle large real-world problems.

In this monograph, finite horizon and infinite horizon discounted Markov decision processes are presented. Along the way the reader is introduced to Markov chains, dynamic programming, and contraction mappings. For the reader who has access to a computer, the various algorithms developed for the resolution of the related optimization problems can be readily programmed and tested.

Specifically, in Chapter 1, the Markov decision process model is defined, and examples are presented in the areas of marketing, machine productivity, and stock control. In Chapter 2 the iterative method of dynamic programming is used to resolve the optimization problem for processes of a limited duration. The theory for infinite horizon processes with future rewards discounted is developed in Chapter 3; and Chapter 4 contains solution techniques for these processes.

As the monograph is intended to be suitable for undergraduate use, the prerequisites are minimal. Except for the last section, Section 4.4, it is assumed only that the reader has some knowledge of sequences and series of real numbers (e.g., the concept of absolute convergence) and some familiarity with matrix algebra (e.g., the inverse and the transpose of a matrix). Section 4.4 is presented for those readers who have already had an introductory course in linear programming.

Various people have made significant contributions to the development and publication of this monograph. First and foremost I wish to acknowledge the efforts of the reviewers of the manuscript. Their careful comments and constructive criticism have been extremely helpful. Also, those students at Boston College who have been subjected to preliminary versions of the manuscript have provided guidance through their questions and responses, and their interest and enthusiasm. And the staff at COMAP, along with my very special typist ML, have been most skillful and most cooperative in moving the material from scrambled notes to printed copy.

1. The Markov Decision Process Model

In this chapter examples involving the sales of a
weekly newspaper, the maintenance of a heavy machine, and
the inventory level of a product will be developed. As
diverse as these examples may seem, they have a common fea-
ture; namely, that each can be modeled as a Markov decision
process. The structure of such models, Markov chains, rewards, and alternatives, will be developed in this chapter
along with the examples. In the subsequent chapters the
associated optimization problems and solution techniques
will be addressed.

1.1 Markov Chains

In many cities there is a weekly newspaper directed at
the college age market--papers that ten years ago would
have been referred to as "underground." These papers fea-
ture movie, theater, and record reviews, along with articles on topics such as energy and environmental problems,
bicycle routes, natural foods, and Monty Python. Advertising rates for such papers would probably be contingent upon
circulation levels. Suppose one publisher uses two rates:
a high rate if the previous week's sales exceeded 50,000
and a low rate otherwise. Examining past records, the publisher can then categorize each week's sales as either high
(above 50,000) or low (less than or equal to 50,000). In
fact, in an attempt to provide some estimate of future
sales, the publisher notes that after a week of high sales,

high or low sales are just as likely the following week, whereas after a week of low sales, a week of high sales follows one quarter of the time. This situation, as we shall see, can be modeled as a Markov chain.

A (finite-state) <u>Markov chain</u> is a process or system with the following properties:

1. At any given time the system is in one of a finite number of states s_1, s_2, ..., s_N;
2. As we move from one time period to the next, that is, through a transition, the system can change states;
3. The probability of the system moving from state s_i to state s_j depends only on the present state s_i and the transition state s_j. This probability is denoted by p_{ij}.

The $N \times N$ matrix $P = [p_{ij}]$ is called the <u>transition matrix</u> of the Markov chain; this matrix completely characterizes the Markov chain.

EXAMPLE 1. For the weekly newspaper, if we assume that the level of each week's circulation depends only on the previous week's circulation, we have an example of a Markov chain. The system has two states, high sales (H) and low sales (L), and transition matrix

$$\begin{array}{cc} & \begin{array}{cc} H & \quad L \end{array} \\ \begin{array}{c} H \\ L \end{array} & \begin{bmatrix} 1/2 & 1/2 \\ 1/4 & 3/4 \end{bmatrix} \end{array}.$$

EXAMPLE 2. Consider the production process in a small stamping plant. In that production centers around a heavy stamping machine, this machine is inspected regularly, say, at the end of each working day. As a result of this inspection and the day's performance, the condition of the machine is then classified into one of three states: s_1, good (the machine is working well); s_2, fair (some breakdowns and minor problems occurring); and s_3, inoperable. Furthermore, this condition of the machine at the end of the day is primarily dependent upon its condition at the beginning of the day and the day's work load. If we assume that each day's work load is the same, then the day-to-day condition of the machine can be considered as a Markov chain. In fact, using past records as a guide, the transition probabilities can be estimated. For this example, the transition matrix might be something like

$$\begin{array}{c} & \begin{array}{ccc} s_1 & \quad s_2 & \quad s_3 \end{array} \\ \begin{array}{c} s_1 \\ s_2 \\ s_3 \end{array} & \left[\begin{array}{ccc} 3/4 & 1/6 & 1/12 \\ 0 & 2/3 & 1/3 \\ 0 & 0 & 1 \end{array}\right]. \end{array}$$

EXAMPLE 3. Consider the monthly inventory level of wood burning stoves at a small alternate energy center. Suppose that the dealer has stock room for at most four; and that in any month it is equally likely that there be 0, 1, 2, 3, or 4 potential buyers. At the end of each month the dealer can reorder stoves from the local supplier, but due to quotas set by the supplier, the dealer must purchase stoves in units of four. Thus, the dealer reorders stoves only if none are on hand. When an order is placed, the supplier provides immediate delivery three quarters of the time. However, one quarter of the time, the supplier's sources are depleted, and the dealer's order must wait a month for reconsideration.

To form a Markov chain, define the state of the process to be the number of stoves on hand at the end of each month. There are five states then: 0, 1, 2, 3, and 4. The transition probabilities for the last four states depend only on the customer probability distribution. For example, suppose the dealer has three stoves on hand at the end of a month. The probabilities of selling 0, 1, or 2 stoves the next month are each 1/5, and so the probabilities of moving to states 3, 2, or 1 are 1/5. The dealer's supply will be exhausted 2/5 of the time (with probability 1/5 that a single potential buyer will be turned away), and so the probability of moving to state 0 is 2/5. Now suppose the dealer has no stoves in stock at the end of the month. Four stoves are ordered from the supplier, and are delivered with probability 3/4. If four are delivered, the number then sold depends on customer demand, as above. For example, the probability of selling three of these four is 1/5, and so the probability of moving from state 0 to state 1 is $(3/4)(1/5) = 3/20$. Similarly the probability of selling all four is 1/5, and so the transition probability from state 0 to state 0 is

$$1/4 + (3/4)(1/5) = 2/5.$$

The complete transition matrix follows:

$$
\begin{array}{c c c c c c}
 & 0 & 1 & 2 & 3 & 4 \\
0 & 2/5 & 3/20 & 3/20 & 3/20 & 3/20 \\
1 & 4/5 & 1/5 & 0 & 0 & 0 \\
2 & 3/5 & 1/5 & 1/5 & 0 & 0 \\
3 & 2/5 & 1/5 & 1/5 & 1/5 & 0 \\
4 & 1/5 & 1/5 & 1/5 & 1/5 & 1/5
\end{array} .
$$

PROBLEM SET 1.1

1. Verify that the transition matrix in Example 3 is correct.

2. Model each of the following as a Markov chain:

   a) The annual yield of a certain apple tree can be categorized as either heavy or light. Assume that the probability of next year's yield being heavy if this year's is heavy is .7, and that the probability of next year's yield being light if this year's is light is .6.

   b) You are betting on red at the Roulette Table (probability of winning is 9/19, and you win or lose the amount you bet). If you have $1, $4, or $5 on hand, you bet $1. If you have $2 or $3 on hand, you bet $2. You start with $4, and quit when you are either broke or have $6. (The associated transition matrix that you construct should have two rows each with a single nonzero entry, and that entry a one.)

   c) Two overhead fixtures provide lighting for a storage room. In any month there is a probability of 1/10 of the lamp in a fixture burning out. A not overly-industrious janitor checks the room's lighting at the end of each month. If there is a single burned out lamp, it is replaced with probability 1/6. If both lamps have burned out, 3/4 of the time both are replaced; otherwise the storage room door is quickly closed and the darkness ignored.

   d) You play the following game. On your first play, you throw a fair die, and the number of spots showing is your point. On subsequent plays, if your previous point was not a one, then you continue to throw the die until you roll a number less than or equal to your previous point. This roll then becomes your point. If your previous point was one, you throw the die once, and that roll is your point.

3. Reconsider the stove inventory example, but with the added condition that if the supplier's sources are depleted when the dealer orders four stoves, the supplier guarantees delivery the next month. (Suggestion: increase the number of states in the associated Markov chain.)

4. Let $P = [p_{ij}]$ be the transition matrix for a Markov chain.

   a) What is the sum of the entries of any one row?

   b) Suppose $p_{ii} = 1$ for some i. Then state s_i is called an <u>ab-</u> <u>sorbing</u> state. Why?

5. a) In the weekly newspaper example, this week's sales are high. Determine the probability that the sales two weeks from now will be high; and the probability that the sales will be low.

   b) As in part a, but assume that present sales are low.

   c) Compare your answers to the above with the entries of the square of the corresponding transition matrix.

6. a) Generalize the results suggested in Problem 5; that is, given that $P = [p_{ij}]$ is the transition matrix for a Markov chain, show that the ij^{th} entry of P^2 is the probability of moving from state s_i to state s_j in two transitions.

   b) Extend this result. Interpret the entries of P^n, where n is a positive integer.

7. Consider the condition of the stamping machine in Example 2.

   a) If the present state is s_2, what is the probability that the machine will be operable, that is, in state s_2, in one day; in two days; in a week; in a year? What is the probability that the machine will eventually become inoperable?

   b) If the present state is s_1, what is the probability that the machine will be in state s_1 in one day; in two days; in a week; in a year?

   c) Regardless of present state, is state s_3 inevitable?

1.2 Rewards

The publisher of the weekly newspaper described in the previous section is interested in total profits from sales and advertising revenue, and so has gathered the following data. Two consecutive weeks of high sales earns 16 units (say, $160 or $1,600) of profit, whereas a week of low sales following a week of high sales nets 10 units. Similarly, high sales after low sales is worth 7 units, and two consecutive low sales weeks, 3 units.

Some obvious questions face the publisher. For example, suppose this week's sales are high. What is the expected income for the next week; or, over the next two weeks? If present sales are low, what are the future expected earnings over the next week, or the next 2, or over the next 52 weeks?

To begin to answer these questions, we use the concept of expected value. If this week's sales are high, there is

-5-

a probability of 1/2 of next week's sales being high and thus a gain of 16, and a probability of 1/2 of low sales and a gain of 10. Hence the expected gain over the next week is $16(1/2) + 10(1/2) = 13$. Similarly if present sales are low, the expected gain for the next week is $7(1/4) + 3(3/4) = 4$.

Suppose we want to calculate the expected earnings for a two-week period. If present sales are high, we have an expected gain of 13 over the first week, and a probability of 1/2 of being in the high sales state the next week and a probability of 1/2 of being in the low sales state. High sales will earn on the average 13 units over the second week, and low sales, 4 units. Thus the total expected gain for the two-week period is

$$13 + 13(1/2) + 4(1/2) = 43/2 = 21\frac{1}{2}.$$

Similarly if present sales are low, the expected gain over the next two-week period is

$$4 + 13(1/4) + 4(3/4) = 41/4 = 10\frac{1}{4}.$$

We can easily extend this technique. For example, if present sales are high the expected gain over the next three-week period is the sum of the expected gain for the first week plus the expected gain for the last two-week period; that is,

$$13 + (43/2)(1/2) + (41/4)(1/2) = 28\frac{7}{8}.$$

If present sales are low, the expected gain over the next three-week period is

$$4 + (43/2)(1/4) + (41/4)(3/4) = 17\frac{1}{16}.$$

Using these data we could calculate the expected gain for a four-week period. And so on, for any period length we wish. Or, better yet, a computer can be programmed to perform these iterative calculations for us.

In general all the optimization problems we will be considering will involve gains, profits, rewards, costs, losses, or whatever. Notice, though, that in working with the gains in the above example, the original profit data for a one-week transition, the 16, 10, 7, and 3, were used only in the calculation of the expected gains for a one-week period. All subsequent calculations involved these expected gains of 13 and 4 and the data of the transition matrix. This will always be the case, and so, in order to reduce the amount of data necessary to define a Markov decision process model, we will work with expected gains. Moreover, we will treat these data as representing profit or reward, and consider the associated maximization problems. The theory developed would apply directly to a model involving the minimization of costs or losses, if we simply record these losses as negative gains.

Thus to each state s_i in a Markov chain, we assume that we can associate or calculate an expected reward or gain, and denote the quantity by q_i. This q_i represents the expected gain earned through one transition if the initial state is s_i. A negative q_i indicates a loss or cost.

EXAMPLE 1. In the weekly newspaper example as described, $q_1 = 13$ and $q_2 = 4$.

EXAMPLE 2. The production level in the stamping plant described in Example 2 of the last section is a function of the condition of the heavy stamping machine. In fact, suppose that the entries in the following array are reasonable estimates on the number of units the plant produces daily.

		State at end of day		
		s_1	s_2	s_3
State at	s_1	500	440	200
beginning	s_2	-	360	150
of day	s_3	-	-	0

Furthermore, suppose that the plant sells the units for $20 each, but is contracted to produce at least 200 a day. Any day production level is less than 200, a penalty fee of $12 for each unit short of 200 is imposed. Letting the q_i's represent expected income, we have

$$q_1 = \$20[500(3/4) + 440(1/6) + 200(1/12] = \$9300$$

$$q_2 = \$20[360(2/3) + 150(1/3)] - \$12(50)(1/3) = \$5600$$

$$q_3 = -\$12(200) = -\$2400.$$

PROBLEM SET 1.2

1. Suppose that in the game described in Problem 2d of Section 1.1, at the end of each play, if your point is a prime number, you win that amount in dollars, and if your point is not prime, you lose that amount. Compute the q_i's.

2. The apple grower of Problem 2a of Section 1.1 nets a profit of
    $20,000 from the orchard if the yield is heavy and the previous
    year's yield was light. However if the previous year's yield was
    also heavy, present demand is less and profit is reduced to
    $19,000. With a light yield, the grower has less to sell, and
    earns only $10,000, regardless of the previous year's yield.
    Suppose this year's yield is heavy. Determine next year's ex-
    pected gain, that is, determine q_1. Similarly, determine q_2.

3. The stove dealer of Example 3 of Section 1.1 sells the stoves for
    $200 apiece, buys them from the supplier for $120 each, and esti-
    mates the cost of the loss of customer good will for each poten-
    tial stove buyer whose needs cannot be met at $40. Assuming that
    these are the only profits and costs involved, calculate the q_i's.

4. a) If the stamping machine of Example 2 of this section is pres-
        ently working well, what is the expected gain over the next
        two-day period?
    b) As above, but suppose the condition of the machine is pres-
        ently in state s_2? in state s_3?
    c) Determine now the expected gain over a three-day period, as-
        suming that the initial state is s_1.

1.3 Alternatives

So far in our examples the decision makers have been
rendered powerless. The newspaper publisher has no influ-
ence over future sales; the stamping plant manager has no
power over a stamping machine destined for inoperability
(see Problem 7 of Section 1.1); and the stove dealer has no
control over the number of potential customers for which
the dealer will have no stoves. We will now remedy this
situation by introducing choices or alternatives into the
models.

EXAMPLE 1. Consider the example of the weekly maga-
zine, first developed in Section 1.1. It is most likely
that the publisher could take actions to increase sales,
such as advertising on radio and television, sponsoring
weekly puzzle contests, or conducting a door-to-door subscription campaign. Such actions would cost money and so
initially would decrease profits, but might pay off in the
long run through higher circulation rates and advertising
revenue. Specifically, let us suppose that if a week's
sales are high, the publisher can take one of two actions:
either do nothing in the way of promotion, or else advertise on weekend radio. Assume that the data already presented in Sections 1.1 and 1.2, the transition probabili-

ties of 1/2 and 1/2 and expected gain of 13, correspond to the no promotion action. On the other hand, let us suppose that advertising on radio increases the probability of the next week's sales being high to 4/5 but costs 4.8 units. Since two consecutive high sales weeks earns 16 units, and a week of low sales after a week of high sales, 10 units (see Section 1.2), the net expected gain for advertising then would be

$$16(4/5) + 10(1/5) - 4.8 = 10.$$

Similarly, if sales are low, suppose the publisher can choose from one of three alternatives, either sponsor no type of promotion, or subsidize a reduction in newsstand price, or promote a door-to-door subscription campaign; and that the relevant data for these alternatives are as follows:

Action	Expected gain	Probability that the next sales state will be	
		high	low
No promotion	4	1/4	3/4
Price reduction	2	1/2	1/2
Subscription campaign	-2	6/7	1/7

In sum, if sales are high, the publisher has a choice of two actions to take and, if sales are low, three actions. Furthermore, the action taken affects both the transition probabilities and the net expected gain. In general, a process such as this we call a <u>Markov decision process</u>. We have a system or process that can be in one of a finite number of states; for each state there is a set of alternatives, representing possible actions available to a decision maker; and for each alternative there are associated transition probabilities and an expected gain.

We will denote the total number of alternatives available if the system is in state s_i by A_i, and will number these alternatives simply as 1, 2, 3, and so on, up to A_i. In general the index k will be used in reference to an alternative, and the corresponding expected gain and transition probabilities will be characterized by use of a superscript. Thus if a system is in state s_i, the expected gain if alternative k is used will be denoted by q_i^k, and the transition probabilities by p_{ij}^k.

EXAMPLE 1 (continued). For the weekly newspaper example, $A_1 = 2$ and $A_2 = 3$. The expected rewards and transition probabilities are given in Table 1.1.

TABLE 1.1

State i	Alternative k	q_i^k	p_{i1}^k	p_{i2}^k
1 (High sales)	1 (no promotion)	13	1/2	1/2
	2 (advertise)	10	4/5	1/5
2 (Low sales)	1 (no promotion)	4	1/4	3/4
	2 (reduce price)	2	1/2	1/2
	3 (subscription campaign)	-2	6/7	1/7

EXAMPLE 2. Consider the operation of the stamping
plant as described in Example 2 of Section 1.1, and expanded upon in Example 2 of Section 1.2. At the end of each
working day the condition of the heavy stamping machine,
the major component in the production process, is classified as either: 1, good; 2, fair; or 3, inoperable. Now
it is realistic to assume that after learning of the status
of the machine, the plant manager can take some form of action to maintain or improve the machine's condition. For
example, the plant may have evening employees available to
perform routine upkeep and maintenance on the machine, and
a team of mechanics ready to be called in to perform a major overhaul if necessary. Also the machine probably could
be replaced, but at a cost in time and money. Specifically, let us assume that if the machine is working well, that
is, is in state 1, then two alternatives are available: 1,
perform routine maintenance; and 2, perform routine maintenance and make minor repairs. If the machine is in state
2, then a major overhaul on the machine may be worthwhile,
and so we have three alternatives: 1, routine maintenance;
2, minor repairs; and 3, major overhaul. If the machine is
inoperable, and so in state 3, it may be advisable to do
nothing if the plant is about to shut down. If a working
machine is required though, then the present machine could
be subjected to a major overhaul, or could be replaced.
Thus in state 3, we have three alternatives: 1, do nothing; 2, major overhaul; and 3, replacement. The relevant
data for these alternatives could be estimated by the
plant's management (probably a rather difficult and

imprecise task), and might look something like the data of Table 1.2, where the q_i^k are in \$100 units:

TABLE 1.2

State i	Alternative k	q_i^k	p_{i1}^k	p_{i2}^k	p_{i3}^k
1	1 (routine maintenance)	93	3/4	1/6	1/12
	2 (minor repairs)	85	5/6	1/9	1/18
2	1 (routine maintenance)	56	0	2/3	1/3
	2 (minor repairs)	48	0	4/5	1/5
	3 (major overhaul)	-10	4/5	1/10	1/10
3	1 (do nothing)	-24	0	0	1
	2 (major overhaul)	-75	2/3	1/6	1/6
	3 (replace machine)	-300	1	0	0

EXAMPLE 3. Consider the inventory problem discussed in Example 3 of Section 1.1 and Problem 3 of Section 1.2. The stove dealer could stock a maximum of four stoves, and could reorder stoves only in units of four. Suppose now that the dealer also has the option of ordering two stoves from the supplier, and so can use this alternative when stock level is two or less. Suppose the supplier can provide immediate delivery of two stoves with probability 4/5, but that in this case, charges the dealer \$150 per stove. Assume as before that when the supplier cannot meet completely an order from the dealer, no transaction between the two takes place and the order is simply cancelled. Using the probability data from the Example of Section 1.1, and the profit and cost data from Problem 3 of Section 1.2, the transition probabilities and expected gains can be calculated for this expanded model, and are given in Table 1.3.

TABLE 1.3

State i (# in stock)	Alternative k	Expected return (in $)	p_{i0}^k	p_{i1}^k	p_{i2}^k	p_{i3}^k	p_{i4}^k
0	1 (order 4)	-80	2/5	3/20	3/20	3/20	3/20
	2 (order 2)	-51.20	17/25	4/25	4/25	0	0
	3 (order 0)	-80	1	0	0	0	0
1	1 (order 0)	112	4/5	1/5	0	0	0
	2 (order 2)	64	12/25	1/5	4/25	4/25	0
2	1 (order 0)	256	3/5	1/5	1/5	0	0
	2 (order 2)	131.20	7/25	1/5	1/5	4/25	4/25
3	1 (order 0)	352	2/5	1/5	1/5	1/5	0
4	1 (order 0)	400	1/5	1/5	1/5	1/5	1/5

Note that in the table a third alternative has been added for when the system is in state 0, namely, of ordering no new stoves. Such an action might be reasonable when the dealer wishes to reduce stock, say, near the end of the heating season. It could be that the model as developed represents the dealer's situation only through the winter months, and that otherwise the dealer has no potential stove buyers and must pay storage costs for any stoves remaining in stock. Such a variation can be easily handled by introducing "terminal values" into the models. See Section 2.1.

To summarize, in this chapter, Markov decision process models have been developed for three types of situations. The first involved the profits of a marketable product, namely, a weekly newspaper; the second, the productivity and maintenance of a machine; and the third, the inventory level of a stock of goods. While the assumptions we have placed on our examples have been rather restrictive, it is hoped that the reader perceives each of the models as being realistic, and appreciates the ease with which each could be expanded upon in order to provide as many states and as many alternatives as necessary to accurately reflect a given situation. Our data has been kept small so that the underlying concepts are more transparent and the calculations to come more manageable. In any application much more data would be involved and a computer would be used to answer the associated optimization problems.

PROBLEM SET 1.3

1. Verify that the data in Table 1.3 follows from the probabilities and gains given in Example 3 of Section 1.1, Problem 3 of Section 1.2, and Example 3 of this section.

2. Calculate the expected returns and transition probabilities for the stove dealer example, using the data as given except for one change, the customer probability distribution. Assume that the probability of 0, 1, 2, 3, and 4 potential customers per month is 1/10, 1/5, 2/5, 1/5, and 1/10, respectively.

3. Reconsider the stove dealer example, under the added assumption that:

   a) the dealer pays a storage fee of $10/month for each stove in stock but not sold; or,
   b) the dealer can also order one stove from the supplier at the end of a month, with the supplier providing immediate delivery with probability 7/8 and charging $160 for the stove, if delivered.

Construct a Markov decision process model for each of the following problems, determining states, transition probabilities, and expected gains. Do not attempt yet to answer the optimization questions raised.

4. You are playing the game described in Problem 2d of Section 1.1 and Problem 1 of Section 1.2, but before each play, you have the option of either using the house's fair die or your own weighted die. Your die is weighted in the 4-5-6 corner, making a roll of 4, 5, or 6 twice as likely as a roll of 1, 2, or 3. (Thus, the probability of rolling either a 1, 2, or 3 is 1/9; and the probability of a 4, 5, or 6 is 2/9.) When should you slip in your weighted die?

5. Suppose the data given in Problem 2a of Section 1.1 and Problem 2 of Section 1.2 correspond to when the apple grower uses only a standard amount of fertilizer, and that early in the growing season the grower also has the option of applying a special but costly fertilizer to the orchard. This process costs $2500, but increases the probability of a heavy yield to .9 if the previous year's yield was heavy, and to .6 if the previous year's yield was light. When should the grower use this fertilizer?

6. The dealer of a certain product sells either 0, 1, or 2 units of the product each week; with a probability of 1/4 of selling 0

units, 1/2 of selling 1, and 1/4 of selling 2. The dealer sells
the units for $24; and can stock at most 5 units at any one time.
There is weekly storage cost of $1/unit for each unit in stock on
Monday and not sold by that Friday. The dealer always wants to
have enough units on hand so as to meet any demand. The dealer
may reorder units from one of two different suppliers at the end
of each week. A local supplier promises immediate delivery for
next week's sales, at a dealer cost of $10/unit, but orders must
be placed in units of two. A distant supplier requires a full
week for delivery, charges the dealer $3/unit plus a $12 shipping
charge independent of the number ordered, and only handles orders
of two or more. What should be the dealer's reordering policy?

7. The degree of sharpness of the cutting blade in a large printing
press is determined every evening by a microgauge. The degree of
sharpness ranges from 1 to 5, with 1 denoting the sharpest level.
It has been found that after one day's operation, the degree of
sharpness of the blade can decrease by at most one unit, with the
probability of the dulling a function of the degree at the beginning of the day. These probabilities are as follows:

initial degree	1	2	3	4
probability of dulling	.05	.07	.10	.14

Paper jams can occur at the cutting section of the press.
The frequency of these jams is a function of the condition of the
blade. Estimates on the total number of jams per day, given the
degree of sharpness of the blade in the morning, are as follows:

initial degree	1	2	3	4	5
number of jams	1	3	5 1/2	8 1/2	12

Each paper jam costs $4 in time and paper lost.

The evening maintenance crew can remove the blade from the
press, sharpen it to degree 1, and return it to the press for the
next day's production. This operation costs $66. When should
the blade be sharpened so that total costs are minimized?

8. A stock broker has a client who wants to sell a number of shares
in a certain stock. The total value of the shares fluctuates
weekly in $100 increments between $2000 and $2400; and the fol-
lowing table lists the probabilities of next week's value, given
this week's value.

		Value in a week				
		2400	2300	2200	2100	2000
Present	$2400	3/5	2/5	0	0	0
value	2300	2/5	2/5	1/5	0	0
	2200	0	3/10	2/5	1/5	1/10
	2100	0	0	1/5	3/5	1/5
	2000	0	0	0	2/5	3/5

The broker estimates that a decision of not to sell costs the client a loss of $10/week in interest. How should the broker advise the client?

9. The annual yield of a farmer's potato crop is affected by the Colorado potato beetle. At the end of the growing season the farmer categorizes the level of that year's infestation of the beetle as zero (Z), light (L), moderate (M), or heavy (H). The farmer has determined that $200 of marketable potatoes are lost when the infestation level is L, $700 when the level is M, and $1500 when the level is H. No losses are sustained when the infestation level is zero.

To combat the beetle the farmer can spray the fields early in the summer with a mild insecticide. This operation costs $800, and must be done before that year's level of infestation can be determined. If last year's level of infestation was zero, then without spraying, it is equiprobable that this year's level will be Z or L, whereas spraying guarantees another year at zero level. If last year's level was L: then without spraying, Z, L, and M levels are equiprobable this year; and with spraying, Z and L levels are equally likely. Following an M level: L, M, and H levels are equiprobable without spraying; and Z, L, and M levels with spraying.

If last year's level was heavy and no spraying is used, M and H levels are equiprobable this year. However, if a heavy infestation follows the use of the insecticide, the beetle develops a partial resistance to the chemical and spraying is of less value. In particular, if last year's level was H and the farmer sprays this year, then, if no insecticide was used last year, all four levels Z, L, M, and H are equally likely this year, whereas if the insecticide was used last year, the levels L, M, and H are equally likely.

When should the farmer use the insecticide so that costs plus losses are minimized?

10. A company provides its executives with a small fleet of automobiles for business use. Each summer a decision is made on how to

implement this service for the ensuing year. Options are to keep the present vehicles, if they are not too old, or to lease or buy new vehicles.

A leasing firm can provide the company the use of new automobiles at an annual rate of $2500 per vehicle. The firm also allows its customers with one-year old leased cars to renew their leases and continue to use these vehicles for a second year, at a reduced annual cost of $2300. Because of reliability and maintenance problems, the leasing firm recalls all leased vehicles after two years of service.

The company can also buy and maintain its own fleet. New automobiles cost $9000, and can be used for at most three years. Annual maintenance costs, the probability of a complete breakdown during any given year, and resale value all depend on the age of the automobiles and are as follows:

Age (years)	0	1	2	3
Annual maintenance costs $	200	400	900	---
Probability of complete breakdown	.03	.06	.12	---
Resale value ($)	---	7200	5700	4500

In the event of a complete breakdown, the automobile must be replaced immediately with a leased, latest-model vehicle. Estimated leasing costs for the fraction of the year that the leased automobile would be used is $1500; and the estimated salvage value of the company lemon, regardless of age, is $1800.

What policy should the company follow so that the transportation needs of its executives are met and costs are minimized?

11. A real estate agent is contacted by up to four prospective home buyers each day, with the probability distribution for the number of buyers as follows:

number	0	1	2	3	4
probability	1/6	1/3	1/4	1/6	1/12

After considering overhead costs, selling prices, brokers fees, the likelihood of making a sale, and so on, the agent has determined that he/she nets on the average $48 for each buyer handled. However, the broker can handle two a day, at most. If more buyers than the broker can handle make contact, up to four buyers can make appointments with the agent for the following day. These buyers definitely return the next day, and their presence does not affect the number of new buyers for that day. However, the agent has found that, under no circumstances, is it worthwhile to schedule more than four appointments for the following day; it is estimated that a buyer not assisted or rescheduled costs $96 in good will.

Each evening the agent notes the number of appointments scheduled for the next day, and then decides on whether to call in a part-time assistant to help the following day. The assistant is paid $12/day plus commission. By considering this commission, the assistant's selling capabilities, and so on, the agent has determined that he/she nets $30 for each buyer the assistant handles. The assistant can handle two buyers a day at most, and the arrangement between the agent and the assistant is that if both are working, the agent handles the first two buyers of the day. When should the agent call in the assistant?

12. A town has two traffic signals at its one major intersection so that traffic safety and flow need not be dependent upon a single light. The Highway Department (HD) has found that, if they do not use any regular inspection and maintenance program, a signal will break down in any one week with probability .2 (and, therefore, both will break down with probability .04). If the HD routinely inspects the signals and changes lamps, or services them in any way during the week, breakdowns occur the next week with probability .1. Now the operation of the signals is also observed weekly by the town sheriff, and the sheriff notifies the HD of any malfunction. If informed that both signals are not operating, the HD performs emergency service and has both working for the beginning of the next week. If only one signal is reported faulty, the HD can again use its emergency service, guaranteeing operation of both for the beginning of the next week, or can employ normal channels to repair the signal. With normal channels the probability that a worker will be available to repair the signal and have both working for the beginning of the next week is .5; and, if the repair is not made, the probability of the second signal at the intersection breaking down is .2. Routine inspection and maintenance costs the town on the average $10/week; emergency service, $60; and service through normal channels, $25. The only other related cost is incurred when both signals are broken down. Then the town must pay its deputy officer $200/week for extra traffic control duty. When both signals are working, should the HD perform routine maintenance; and if one signal is reported faulty, should they use the emergency repair service or normal channels? (Let the states of the system be the number of signals reported faulty during a week's operation.)

2. Markov Decision Processes: Finite Horizon

Having provided the decision makers of our examples with various alternative actions to select from, we now consider the general problem of determining the best course of action to follow. In this chapter, we will assume that the number of transitions remaining in the process is finite, and will seek a policy that optimizes the total expected reward over this finite horizon. In the first section, policies, optimal policies, and optimal policy values will be discussed. In the second section, an iterative process for calculating optimal policies and values will be developed. In Chapters 3 and 4, optimal policy questions will be considered for Markov decision processes with an unlimited number of transitions remaining, that is, for processes with no anticipated termination.

2.1 Optimal Policies and Values

It could be that the weekly newspaper described in Chapter 1 is published only from September through May, and that publication is suspended during the summer months while the college market is elsewhere. In fact, suppose that only three more weeks remain in the present publishing season, and that past week's sales were high. The publisher is faced with a decision, as described in Example 1 of Section 1.3; namely, should money be invested on weekend radio advertising? Upon reflection the publisher realizes that since this decision affects the probability of high and low sales for the remaining weeks, an intelligent choice can be made now only after the potential profits for the final two weeks of operation are considered. Similarly a determination of profit for the final two weeks is contingent upon the decisions made when both two weeks and one week remain. This suggests that in order to determine the decision to make now, the publisher must consider all possible courses of action that can be taken over the remaining three weeks, determine the total expected profits from each, and then select a course of action that has the maximal expected gain.

An example of a course of action for the publisher to follow would be to use the no promotion alternative now and, regardless of circulation levels in subsequent weeks, continue to use the no promotion alternative. Another example would be to use Alternative 2 now (as defined in Table 1.1 of Section 1.3), Alternatives 2 and 3 if in States 1 and 2, respectively, when two weeks remain, and Alternatives 1 and 2 if in States 1 and 2, respectively, for the final week. In general we will call such a prescribed course of action a <u>policy</u>. Thus, for a given number of transitions remaining in the process and a given initial state of the system, a policy prescribes the alternatives to be used through the remaining transitions.

Policies may be varied. A policy description of the alternative to use at a given point in the process depends on the state of the system, and may also depend on the number of steps still remaining and the past history of the process. Nevertheless, for a Markov decision process with a finite horizon, given a fixed initial state and number of remaining transitions, there are only a finite number of policies. For example, for the newspaper publisher's situation, there are 72 available policies. One of two alternatives must be selected immediately; then after one transition, sales can be either high or low, and so one of two alternatives must be prescribed in case sales are high, and one of three in case sales are low. Similarly the alternatives to use after two transitions must be listed. Thus we have $2 \times 2 \times 3 \times 2 \times 3 = 72$ policies.

The total expected gain for a given policy can be determined by working backwards, starting with the expected gains with one transition remaining, the q_i^k's; and then using the transition probabilities and these q_i^k's to determine the expected gains with two transitions remaining; and so on. For example, for the publisher's situation as described, with three weeks remaining and present sales high, the policy of always using the first alternative, the no promotion alternative, out of each state, has a total expected gain of 28 7/8, as determined in Section 1.2. The total expected gain of a policy will be called its _value_.

For a Markov decision process with n transitions remaining and present state s_i, a policy that delivers the maximal value will be called an _optimal policy_, and its value will be denoted by $v_i(n)$. Note that for a fixed n and i, the value of an optimal policy, the $v_i(n)$, is unique, but that there may be more than one policy with value equal to this maximal value.

Now one way to determine an optimal policy would be to list all possible policies, calculate the value of each, and then select one whose value is equal to the largest number in this set. However there is a much more efficient method. It is an inductive approach, based on what is called the Principle of Optimality from Dynamic Programming (Bellman[1]). In our terms, this principle states that an optimal policy for a process with n transitions remaining and present state s_i must make use of optimal policies for the system with (n-1) - steps remaining; that is, if after the initial decision and transition the system is in state s_j, the original optimal policy must now constitute an optimal policy for the system with initial state s_j and (n-1) - steps remaining. This principle is intuitively clear; if it were not true for a policy assumed optimal, then this policy would in fact not be optimal but instead could be improved upon over the last (n-1) - transitions.

Consider now the quantity $v_i(n)$, the value of an optimal policy for the process with the present state s_i and n transitions remaining. Such a policy prescribes the use of one of A_i alternatives out of state s_i on the first transition. Suppose alternative k is prescribed. The expected gain on the initial transitions would be q_i^k, and the probability of moving to state s_j would be p_{ij}^k. If the system does in fact move to state s_j, then, from the Principle of Optimality, the total expected gain for the optimal policy over the last (n-1) - transitions would be $v_j(n-1)$. Hence the total expected gain for the n transitions would be

$$q_i^k + \sum_{j=1}^{N} p_{ij}^k \, v_j(n-1)$$

where N is again the number of states in the system. It follows that $v_i(n)$ satisfies the recursive relationship

$$(1) \qquad v_i(n) = \underset{1 \leq k \leq A_i}{\text{Maximum}} \{q_i^k + \sum_{j=1}^{N} p_{ij}^k \, v_j(n-1)\}.$$

This is the crucial relationship for a Markov decision process over a finite horizon. It yields immediately the inductive approach that we will use in the next section to determine optimal policies and values. Moreover in Chapter 3 we will encounter an analogous relationship for Markov decision processes with infinite horizons.

One final note. So far in discussing processes with a finite number of transitions remaining, we have yet to consider whether the final state of the system has any affect on the total expected gain. It could be that it does not. In the weekly newspaper example, the circulation level for the last week's sales in June would affect the earnings over that last transition, but probably nothing more. On the other hand, the states of our process may correspond to the condition, and thus the value, of an object, such as a machine or automobile, that we wish to sell at the end of the n transitions, and that the income we receive from this sale is to be considered as part of the total expected gain. Or, as mentioned at the end of Example 3 in Section 1.3, our stove dealer model may correspond to the dealer's situation through the winter months only, and that stoves remaining in inventory over the summer months represent real costs to the dealer in terms of storage fees, unavailable capital, and depreciation.

So as to be able to incorporate these considerations into our models, we define the terminal value of state s_i to be the additional reward earned if the system terminates in s_i; and we denote this quantity by $v_i(0)$. As will be seen in the next section, these data are very easily incorporated into the solution process that we develop. For the present, the reader may note that with these terminal values and notation, the above relationship (1) is now valid for any $n \geq 1$.

EXAMPLE 1. Assuming that in the weekly newspaper example, the level of the last week's sales affects only the income earned over the last transition, we have terminal values

$$v_1(0) = v_2(0) = 0.$$

EXAMPLE 2. After deducting costs, suppose that the stove dealer of Example 3 of Section 1.3 estimates that each stove left in stock over the summer months is worth only \$25. Since in that example state i corresponds to having i stoves in stock, we have

$$v_i(0) = \$25i \text{ for } i = 0, 1, 2, 3, 4.$$

PROBLEM SET 2.1

1. How many policies are available to the weekly newspaper publisher if present sales are low and

   a) 3 weeks remain;
   b) 13 weeks remain.

2. Suppose the stove dealer of Example 3 of Section 1.3 sells stoves from September through the following April and that two stoves remain in stock from last season's sales. How many policies are available to the dealer for the coming heating season?

3. The stock broker of Problem 8 of Section 1.3 is given this week plus two additional weeks to sell the client's stock. This week the market value of the stock is \$2100.

   a) How many possible policies are there for the broker?
   b) Determine terminal values for use in the event that the decision made each week is to not sell.

4. The real estate agent of Problem 11 of Section 1.3 works a six-day week, from Mondays through Saturdays. Moreover, no appointments are made over the weekend; any buyer on Saturday that cannot be handled is considered lost. Determine the terminal values that would apply on Saturday evenings.

5. a) Suppose $c_1, \ldots, c_m$ are any real numbers, and $p_1, \ldots, p_m$ are real numbers between 0 and 1 such that

   $$\sum_{k=1}^{m} p_k = 1.$$

   Then

   $$c = \sum_{k=1}^{m} p_k c_k$$

   is a weighted average of the c_k's. Show that

   $$c \leq \text{Max}_{k} \{c_1, \ldots, c_m\}.$$

   b) Suppose one transition remains in a Markov decision process presently in state s_i. Show that the value of a stochastic or mixed policy (for example, a policy of the form – use alternative k with probability p_k) can never exceed $v_i(1)$. (Hint: Use part a, letting $m = A_i$ and

-22-

$$c_k = q_i^k + \sum_{j=1}^{N} p_{ij}^k v_j(0).)$$

c) Extend this result. Show that for the problem of maximizing the total expected gain of a Markov decision process over a finite horizon, stochastic policies need not be considered.

6. True or false: If all the gains for a process are positive, then for each i, the function $v_i(n)$ is strictly increasing in n.

2.2 Determination of Optimal Policies and Values

Last section we defined $v_i(n)$ to be the total expected gain for a Markov decision process with n transitions remaining and present state s_i, and we showed that these values satisfy the relationship

$$(1) \qquad v_i(n) = \underset{1 \le k \le A_i}{\text{Maximum}} \{q_i^k + \sum_{j=1}^{N} p_{ij}^k v_j(n-1)\}.$$

This relationship implies that if we know the $v_j(n-1)$ for all j between 1 and N, determination of $v_i(n)$ is immediate. We would simply calculate the A_i quantities

$$q_i^k + \sum_{j} p_{ij}^k v_j(n-1)$$

and set $v_i(n)$ equal to the largest; and an alternative k at which this maximum is attained would be an alternative an optimal policy would use for the initial transition out of state s_i. Similarly, in order to determine the $v_j(n-1)$'s, all we need to know is the $v_j(n-2)$'s. And so on. This suggests that in order to determine $v_i(n)$ and an associated optimal policy, we work backwards, and first calculate the $v_j(1)$'s, then the $v_j(2)$'s, and so on. After n iterations we will have $v_i(n)$, and the listing of the alternatives for each step and state at which the corresponding maximum was attained would provide an optimal policy. We demonstrate with two examples.

EXAMPLE 1. Consider the problem raised in the last section. The weekly newspaper publisher has three weeks remaining in the present season, and last week's sales were high. What policy should be followed?

Before we begin the calculations we repeat the data for the model, appending Table 1.1 of Section 1.3 with the terminal values discussed in Example 1 of Section 2.1.

TABLE 2.1

State i	Alt k	q_i^k	p_{i1}^k	p_{i2}^k	$v_i(0)$
1	1	13	1/2	1/2	0
	2	10	4/5	1/5	
2	1	4	1/4	3/4	0
	2	2	1/2	1/2	
	3	-2	6/7	1/7	

We must determine $v_1(3)$ and a corresponding policy with this as its value. Consider first $v_1(1)$ and $v_2(1)$. Since the terminal values are zero,

$$v_1(1) = \underset{k=1,2}{\text{Max}} \{q_1^k\} = \text{Max} \{13, 10\} = 13$$

and

$$v_2(1) = \text{Max} \{4, 2, -2\} = 4.$$

Thus when one week remains, if sales are high, the publisher uses Alternative 1 with an expected gain of 13; if sales are low, Alternative 1 with gain of 4.

Using these numbers and data from the Table, we have from (1):

$$v_1(2) = \underset{k=1,2}{\text{Max}} \{q_1^k + \sum_{j=1}^{2} p_{1j}^k v_j(1)\}$$

$$= \text{Max}\{13+13(1/2)+4(1/2), \ 10+13(4/5)+4(1/5)\}$$

$$= \text{Max}\{21.5, \ 21.2\} = 21.5 \text{ when } k = 1.$$

$$v_2(2) = \underset{1 \leq k \leq 3}{\text{Max}} \{q_2^k + \sum_{j=1}^{2} p_{2j}^k v_j(1)\}$$

$$= \text{Max}\{4+13(1/4)+4(3/4), \ 2+13(1/2)+4(1/2),$$
$$-2+13(6/7)+4(1/7)\}$$

$$= \text{Max}\{10.25, \ 10.5, \ 9.7143\} = 10.5 \text{ when } k = 2.$$

Hence

$$v_1(3) = \underset{k=1,2}{\text{Max}} \{q_1^k + \sum_{j=1}^{2} p_{1j}^k v_j(2)\}$$

$$= \text{Max}\{13+21.5(1/2)+10.5(1/2),$$
$$10+21.5(4/5)+10.5(1/5)\}$$

$$= \text{Max}\{29, \ 29.3\} = 29.3 \text{ when } k = 2.$$

Thus with three weeks remaining and present sales high, the publisher has a total expected gain of 29.3 units. The optimal policy that attains this value is to use Alternative 2, to advertise on radio, the first week; when two weeks remain, if sales are high, use Alternative 1, the no promotion alternative, and if sales are low, Alternative 2, the reduction in price alternative; and when one week remains, use Alternative 1 regardless of sales level.

We summarize these results in the following table, where the calculations have been carried through $n = 6$. The reader should be able to verify the data of the table. (The notation $a_i(n)$ appears in the table. We use $a_i(n)$ to denote the alternative to use out of state s_i in an optimal policy when n transitions remain.)

TABLE 2.2

n	1	2	3	4	5	6
$v_1(n)$	13	21.5	29.3	37.04	44.7692	52.4989
$v_2(n)$	4	10.5	18	25.6857	33.4180	41.1476
$a_1(n)$	1	1	2	2	2	2
$a_2(n)$	1	2	2	3	3	3

The table lists optimal values and optimal policies for the process when six or fewer weeks remain. Notice how the optimal alternative out of state s_2 changes with n. When four to six weeks remain the costly Alternative 3 is used; evidently the high probability of moving to state s_1 along with the three or more weeks remaining to earn income compensate for the alternative's initial cost. However when only two or three weeks remain to earn income, Alternative 2 with its more immediate gain prevails; and when one week is left, the immediate gain is the only concern.

EXAMPLE 2. Consider the inventory problem of the stove dealer of Example 3 of Section 1.3, working with a finite time horizon and terminal value of $25/stove as suggested in Example 2 of Section 2.1. Using these terminal values and the data of Table 1.3 of Section 1.3, we have for $v_0(1)$, the value of an optimal policy when one month remains and zero stoves are in stock:

$$v_0(1) = \max_{1 \le k \le 3} \{q_0^k + \sum_{j=1}^{5} p_{0j}^k \, v_j(0)\}$$

$$= \max\{-80+25(3/20)+50(3/20)+75(3/20)+100(3/20),$$
$$-51.20+25(4/25)+50(4/25), \, -80\}$$

$$= \max\{-42.50, \, -39.20, \, -80\} = -39.20 \text{ when } k = 2.$$

Similarly

$$v_1(1) = \max\{112+25(1/5), \, 64+25(1/5)+50(4/25)+75(4/25)\}$$

$$= \max[117, \, 89] = 117 \text{ when } k = 1.$$

Table 2.3 lists the results of these calculations extending through n = 4:

TABLE 2.3

n	1	2	3	4
$v_0(n)$	-39.20	87.32	199.50	305.50
$v_1(n)$	117	173.06	271.93	377.74
$v_2(n)$	271	330.94	436.69	541.48
$v_3(n)$	382	490.32	585.79	690.68
$v_4(n)$	450	636.16	743.56	847.50
$a_0(n)$	2	1	1	1
$a_1(n)$	1	2	2	2
$a_2(n)$	1	2	2	2
$a_3(n)$	1	1	1	1
$a_4(n)$	1	1	1	1

Thus when one month remains in the heating season the dealer should order two stoves if none are in stock, and otherwise order none. When two to four months remain, the dealer should reorder four stoves when stock level is zero, and two stoves when stock level is one or two. In fact, it would appear that this reordering policy is probably optimal for any $n \ge 2$; that is, that the $a_i(n)$'s are fixed at these associated values for any $n \ge 2$. (See Problems 8-10 for a consideration of the general question of convergence of the $a_i(n)$'s as n grows large.)

1. Determine optimal policies and values for n = 3 for the Markov decision process with the following data:

State	Alt	q_i^k	p_{i1}^k	p_{i2}^k	$v_i(0)$
1	1	6	.8	.2	
	2	8	.5	.5	2
	3	5	0	1	
2	1	4	.9	.1	
	2	2	.5	.5	10

2. For the Markov decision process given by the following data

State	Alt	q_i^k	p_{i1}^k	p_{i2}^k	$v_i(0)$
1	1	16	1/2	1/2	
	2	12	3/4	1/4	0
2	1	0	1/4	3/4	
	2	-4	3/4	1/4	8

a) Determine optimal values and policies for n = 4.

b) Show that in part a $a_1(4)$ is not unique. (Note in fact that because of this, the value of $a_1(n)$ could change twice as n goes from 1 to 4.)

c) Show that if $v_1(n) - v_2(n) = 16$, then

$a_1(n+1) = 1$ or 2,

$a_2(n+1) = 2$, and

$v_1(n+1) - v_2(n+1) = 16$.

d) Determine $a_1(n)$ and $a_2(n)$ for all n.

3. Consider the weekly newspaper example, with data of Tables 2.1 and 2.2

a) Show that if $v_1(n) - v_2(n) > 10$, then $a_1(n+1) = 2$.

b) Show that if $v_1(n) - v_2(n) > 11.2$, then $a_2(n+1) = 3$.

c) Show that if $11.2 < v_1(n) - v_2(n) < 14$, then $11.2 < v_1(n+1) - v_2(n+1) < 14$.

d) It is the first week of September, the beginning of the 39-week publishing season. What policy should the publisher follow?

4. The stove dealer of Example 2 suspects that the $25 estimate on the value of each stove left in stock over the summer months is too low, and wonders how this value affects the final month's ordering policy.

   a) Show that if this estimate T is greater than $28.24, then $a_0(1) = 1$.
   b) Show that if $T > \$60$, then $a_1(1) = 2$.
   c) Determine T_0 such that if $T > T_0$, then $a_2(1) = 2$.

5. a) The apple grower of Problem 5 of Section 1.3 has leased the orchard for three years. Last year's yield was light. What are the grower's expected earnings for the three years, and when should the special fertilizer be used?

   b) As above, but suppose now that the grower owns the orchard, and plans on selling it in three years. If the final year's yield is heavy, the orchard will sell for $2500 more than if the yield the final year is light.

6. The stock broker of Problem 8 of Section 1.3 has been given five weeks to sell the stock. How would you advise the broker?

7. Regardless of condition, every Saturday the cutting blade in the printing press of Problem 7 of Section 1.3 is sharpened to degree 1 as part of the weekend maintenance plan. What sharpening policy should the evening maintenance crew follow during the five day work week?

8. It was seen in Problems 2 and 3 above that for the associated Markov decision processes, for fixed i, the $a_i(n)$'s converged as n grew large, and it might seem that in general, for any process there should exist an n_0 such that, for each i, $a_i(n) = a_i(n_0)$ for $n \geq n_0$. However this is not the case. Consider the Markov decision process with data as follows:

State	Alt	q_i^k	p_{i1}^k	p_{i2}^k	$v_0(i)$
1	1	8	1/4	3/4	0
	2	6	3/4	1/4	
2	1	1	1	0	0

   a) To start, determine optimal values and policies for n = 4.
   b) Now generalize. Show that if $v_1(n) - v_2(n) < 4$, then $a_1(n+1) = 1$ and $v_1(n+1) - v_2(n+1) > 4$.
   c) Show that if $v_1(n) - v_2(n) > 4$, then $a_1(n+1) = 2$ and $v_1(n+1) - v_2(n+1) < 4$.

d) Conclusion:

$$a_1(n) = \begin{cases} 1, & n \text{ odd} \\ 2, & n \text{ even.} \end{cases}$$

9. For the process

State	Alt	q_i^k	p_{i1}^k	p_{i2}^k	$v_0(i)$
1	1	10	.8	.2	
	2	12	.4	.6	0
2	1	4	1	0	0

Show that

$$a_1(n) = \begin{cases} 2, & n \text{ odd} \\ 1, & n \text{ even.} \end{cases}$$

10. a) Determine, for any n, $v_2(n)$, $v_3(n)$, $v_1(n)$, and $a_1(n)$, for the following Markov decision process:

State	Alt	q_i^k	p_{i1}^k	p_{i2}^k	p_{i3}^k	$v_i(0)$
1	1	1	1	0	0	
	2	0	0	1	0	0
2	1	3	0	0	1	0
3	1	0	0	1	0	0

b) As above, but assume now that $v_1(0) = 5$.

c) Suppose n_0 is a fixed positive integer. Determine $v_1(0)$ for the above process so that $a_1(n)$ is constant for $n \leq n_0$ but is not convergent as $n \to \infty$.

11. Suppose that for a process with only two states, there is an integer n_0 and a constant c such that

$$v_1(n_0) - v_2(n_0) = v_1(n_0+1) - v_2(n_0+1) = c.$$

Prove that for all $n \geq n_0$,

$$v_1(n) - v_2(n) = c.$$

12. Program on a computer the solution process developed in this section. Use your program to determine when the real estate agent of Problem 11 of Section 1.3 and Problem 4 of Section 2.1 should call in an assistant.

3. Markov Decision Processes: Infinite Horizon

We now consider the problem of defining and determining optimal policies and values for Markov decision processes that have no planned termination. Since we have satisfactorily resolved these questions for processes with any finite number n of transitions remaining, it might seem that one way to define an optimal policy for the infinite horizon case would be to consider the optimal policies determined using the iterative technique of the last chapter for n very large. But the examples in Problems 8-10 of Section 2.2 shows that the alternatives corresponding to optimal policies need not become fixed as n grows large; and even if for a given process these alternative values do converge, how do we know this, and how do we know how large n need be when determining the limiting values?

Thus we face anew for processes with an infinite horizon the questions of evaluating policies and determining optimal choices. However, one problem immediately surfaces: the concept of total expected reward that served as the foundation for policy evaluation in the last chapter does not directly extend. For example, for a Markov decision process with all the expected rewards q_i^k positive, no matter what policy we use, the sum of the gains must diverge to infinity as n grows large. Thus other measures for policy evaluation must be developed.

As we shall see one way to assure finite policy values is to discount future rewards. This is the approach that we will use; developing the theory in this chapter and the related solution techniques in the next. Reference to other methods of policy evaluation is made in the Bibliographic Notes.

The major result of this chapter is the Stationary Policy Theorem, stated and proved in Section 3.3. This theorem will show that for processes with an infinite horizon and future rewards discounted, there always exists an optimal policy, and in fact, an optimal policy that is stationary in form (that is, the alternative to use depends only on the state the system is in).

3.1 Discounting Future Rewards

Which would you prefer, $100 a year for the rest of your life, or a flat grant of $1529 now? One hundred dollars a year for 50 years would total $5000, considerably more than $1529, but then the flat grant would provide a sizable sum immediately. In fact, given $1529, we could extract $100 for immediate use, and invest the remainder. At an annual interest rate of 7%, the $1429 would yield $100.03 yearly in interest. Thus, assuming that such a rate of interest is and always will be available, the two plans can be considered equivalent.

This equivalence demonstrates that the present value of future earnings need not be their face value. For example, the $100 a year plan would provide us with $100 in one year. But $93.46 invested now at a 7% annual interest rate would be worth $93.46(1.07) = $100 in a year. Thus $100 one year from now should have present value of $93.46; and similarly the present value of $100 to be provided in two years should be reduced proportionately, and so on.

For the Markov decision processes that we consider in this and the next chapter, we assume that future earnings should be discounted, and we will work with the present value of policies with this reduction of future rewards taken into account. Certainly if the transition period for a Markov decision process model is a year or a month, this would seem to be a realistic assumption. Even if the transition period is such that only a very small discount factor is reasonable, the theory that we will develop will be valid as long as some factor less than one is used in determining the present value of future earnings.

To determine the discount factor in terms of an interest rate, suppose we have an interest rate of r, that is, 100r%, per transition period. Then $M now would be worth $(1+r)(\$M) = \$N$ after one transition, and so the present

value of \$N earned after one transition is \$N/(1+r). Thus
rewards to be earned one transition period into the future
should be discounted by the factor $\alpha = 1/(1+r)$. For example, a 7% annual interest rate would correspond to a discount factor of $\alpha = 1/(1+.07) \simeq .9346$ if the length of the
associated transition period is a year; an annual rate of
10% and transition period length of a month to a factor of
$\alpha = 1/(1+.1/12) \simeq .9917$.

If the discount factor is α, then a reward of M to be
earned two transition periods into the future would have
value αM one period from now, and so would have present
value $\alpha^2 M$. Similarly a reward of M to be earned n periods
into the future would have present value $\alpha^n M$. Thus the
present value of \$100 a year for life, with discount factor
α, would be the dollar sum

$$100 + 100\alpha + 100\alpha^2 + \ldots .$$

Summing to infinity (a long life), we have

$$\sum_{n=0}^{\infty} 100\alpha^n = 100 \sum_{n=0}^{\infty} \alpha^n = \frac{100}{1-\alpha}.$$

(We use the fact that the limit of the geometric series

$$\sum_{n=0}^{\infty} \alpha^n \text{ is } \frac{1}{1-\alpha}, \text{ for any } \alpha, |\alpha| < 1.)$$

Now for $\alpha = .9346$, $100/(1-\alpha) = 1529.05$, as expected from
the above discussion.

Discount factors could be used in Markov decision process models with a finite horizon as well as in models with
an infinite horizon. For example the stove dealer of Example 3 of Section 1.3, working with a six or eight month
selling period, may wish to combine into the model the fact
that earnings gained through immediate sales can be invested in some income producing manner over the remaining
months of the model life and so are more valuable than future earnings.

In general only minor modifications are necessary in
order to incorporate the discounting of future rewards into
the finite horizon model. Suppose the discount factor is
α. Then this factor α would be used in the calculation of
the value of each policy, and $v_i(n)$ now would correspond to
the suitably discounted value of an optimal policy. Since
the present value of the quantity $v_j(n-1)$ earned one transition period into the future is $\alpha v_j(n-1)$, the relationship (1) of Sections 2.1 and 2.2 becomes

(1) $$v_i(n) = \max_{1 \leq k \leq A_i} \{q_i^k + \alpha \sum_{j=1}^N p_{ij}^k v_j(n-1)\}.$$

To determine optimal policies and values then, we would proceed just as in Section 2.2, but now use this relationship between the optimal values.

EXAMPLE 1. For the Markov decision process with data given in Table 2.1 of Section 2.2, we have, for $\alpha = .95$ and $\alpha = .9$,

TABLE 3.1

$\alpha = .95$

n	1	2	3	4	5	6
$v_1(n)$	13	21.075	27.9313	34.4190	40.5751	46.4213
$v_2(n)$	4	10.075	16.7963	23.2456	29.3907	35.2338
$a_1(n)$	1	1	2	2	2	2
$a_2(n)$	1	2	2	2	2	2

TABLE 3.2

$\alpha = .9$

n	1	2	3	4	5	6
$v_1(n)$	13	20.65	26.635	32.0215	36.8694	41.2324
$v_2(n)$	4	9.65	15.635	21.0215	25.8694	30.2324
$a_1(n)$	1	1	1	1	1	1
$a_2(n)$	1	2	2	2	2	2

Contrast these results with the results for the non-discounted case given in Table 2.2 of Section 2.2. With $\alpha = .95$, Alternative 3 out of state s_2 no longer appears in any optimal policy; the promised future earnings, now discounted, no longer compensate for the high initial cost $q_2^3 = -2$. Similarly, with $\alpha = .9$, out of state s_1 only Alternative 1 is used; its higher immediate gain of 13 dominates Alternative 2 and its greater probability of remaining in state s_1.

1. Determine optimal values and policies for n = 3 for the Markov decision process of Problem 1 of Section 2.2, using a discount factor of α = .75.

2. Determine optimal values and policies for n = 4 for the Markov decision process of Problem 2 of Section 2.2, using a discount factor of α = 1/2. Compare your results to the results in the nondiscounted case.

3. Consider the Markov decision process given by the following data:

State	Alt	q_i^k	p_{i1}^k	p_{i2}^k	$v_i(0)$
1	1	1	1	0	0
	2	0	0	1	
2	1	3	0	1	0

a) Show that with no discounting $v_1(2) = 3$ and $a_1(2) = 2$.

b) Alternative 2 out of s_1 has no immediate gain, but guarantees a gain of 3 one transition removed. Alternative 1 has an immediate gain of 1. It would seem then that for a high discount rate, $a_1(2)$ might equal 1. In fact, determine α_0 such that, using discount factor α, we have $a_1(2) = 1$ for $\alpha < \alpha_0$ and $a_1(2) = 2$ for $\alpha > \alpha_0$.

4. For the Markov decision process defined by

State	Alt	q_i^k	p_{i1}^k	p_{i2}^k	$v_i(0)$
1	1	2	1	0	0
	2	1	1/3	2/3	
	3	0	0	1	
2	1	6	0	1	0

and with a discount factor α, determine α_1 and α_2 such that

$$a_1(2) = \begin{cases} 1, & \alpha < \alpha_1 \\ 2, & \alpha_1 < \alpha < \alpha_2 \\ 3, & \alpha_2 < \alpha. \end{cases}$$

5. a) Let $0 < r < 1$ and $\alpha = 1/(1+r)$. Show that

$$1 - r < \alpha < 1 - r + r^2.$$

(Hint: first prove that $\alpha = 1 - r + r^2/(1+r)$.)

b) Conclusion: For a low interest rate r, the associated discount factor α can be approximated by $1 - r$.

6. Suppose that the publisher of the weekly newspaper (Example 1 of Section 1.3 and Example 1 of Section 2.2) does not believe in discounting future earnings, but still believes that the newspaper's future is tentative because the printing press on which the newspaper is run off is rather old. In fact, suppose the publisher estimates that for any given week, with probability β the press will self-destruct and he/she will be out of business for the remainder of the season. How can this contingency be incorporated into the model?

3.2   The Value of a Policy

In order to define an optimal course of action for a Markov decision process with an infinite horizon, let us try to proceed as in the finite horizon case; that is, list and evaluate all possible policies, and then select one that delivers the maximum value.

Suppose our process is presently in state s_i. Then by a <u>policy</u> with initial state s_i we mean, as before, a prescribed rule of action that directs our choice of alternatives to use in the present and in all future transitions. Thus for any number n of steps from the present, and for any state s_j that we might then be in, a policy must prescribe the choice of alternative to use. This choice can depend on n; and since n is unbounded in the infinite horizon case, there are now an infinite number of policies with initial state s_i.

Assuming that future rewards are discounted by the factor α, $0 < \alpha < 1$, we now define the total expected reward for a policy, say D, with initial state s_i. First the policy D prescribes an alternative, say k*, for the transition out of state s_i. The probability of being in state s_j after one transition is p_{ij}^{k*}; denote this probability by $p(1,j)$. Now, using the alternatives directed by D, we could calculate the probabilities of being in the various states after two transitions; denote these probabilities by $p(2,j)$, $1 \leq j \leq N$. Similarly we could calculate the probability $p(n,j)$ of being in state s_j after n transitions. Denoting by $D(n,j)$ the alternative that D prescribes for use after n transitions if the system is in state s_j, the expected reward to be earned by D on the (n+1) - transition would be

$$\sum_{j=1}^{N} p(n,j) \; q_j^{D(n,j)}$$

and the present value of this sum is

$$\alpha^n \sum_{j=1}^{N} p(n,j) \; q_j^{D(n,j)}.$$

Thus the total expected reward of D should be the sum

(1) $$q_i^{k^*} + \sum_{n=1}^{\infty} \alpha^n \sum_{j=1}^{N} p(n,j) \; q_j^{D(n,j)}.$$

One obvious question: Does this sum converge? We shall show that in fact the series in (1) is absolutely convergent. Choose Q such that $|q_i^k| \leq Q$ for all $1 \leq i \leq N$ and $1 \leq k \leq A_i$. Then

$$\left| \sum_{j=1}^{N} p(n,j) \; q_j^{D(n,j)} \right| \leq \sum_{j=1}^{N} p(n,j) \left| q_j^{D(n,j)} \right|$$

$$\leq Q \sum_{j=1}^{N} p(n,j) = Q \cdot 1 = Q.$$

Thus the series of absolute values of the terms of (1) is bounded by

$$Q + \sum_{n=1}^{\infty} \alpha^n Q = \frac{Q}{1-\alpha}.$$

Hence the series in (1) is absolutely convergent, and therefore convergent. We call this sum the <u>value</u> of the policy D, and denote this value by $w_i(D)$. In general, the symbol w will be used for policy values, and the subscript on the w will correspond to the starting state.

Thus, as long as we incorporate into the process a discount factor $\alpha < 1$, to each policy we can associate a finite real number representing the value of the policy. It is these values that we will build upon in the next section. For the present we prove a lemma that we will need relating these values.

Let D be a policy with initial state s_i and prescribing alternative k* on the first transition. If $p_{ij}^{k^*} \geq 0$, it is possible that the system be in state s_j after one transition and so D must contain directions for alternative selections for all future n out of state s_j. This list of alternative choices can be considered to be a policy with initial state s_j; we call the policy a <u>one-step-removed policy</u>, and denote it by $\bar{D}_j$.

<u>LEMMA.</u> Let D be a policy with initial state s_i; and prescribing alternative k* on the first transition out of s_i. Then

(2) $\qquad w_i(D) = q_i^{k*} + \alpha \sum_{j=1}^{N} p_{ij}^{k*} w_j(\bar{D}_j).$

PROOF. If after one step the process is in state s_j, the expected reward to be earned henceforth is $w_j(\bar{D}_j)$; and that has present value $\alpha w_j(\bar{D}_j)$. Since the probability of being in s_j after one step is p_{ij}^{k*}, the present total expected reward to be earned after the first transition is

$$\alpha \sum_{j=1}^{N} p_{ij}^{k*} w_j(\bar{D}_j).$$

Equation (2) follows. (Actually we are using here the fact that for convergent series the limit of a sum is the sum of the limits. See Problem 4.) ###

This lemma allows us to compute policy values for policies of a special form.

EXAMPLE 1. Consider the Markov decision process of Table 2.1 of Section 2.2 as having an infinite horizon with discount factor .9. Let D be the general policy of using Alternative 2 when in State 1 and Alternative 1 when in State 2. (Such a policy is called a stationary policy: the choice of alternative depends only on the state the system is in and is independent of n. For such a policy we denote the alternative to use if in state s_j by D(j). Thus D(1) = 2 and D(2) = 1 for this stationary policy D.) Two values are associated with using policy D: if the initial state of the system is s_1, the total expected gain is the value $w_1(D)$; if the initial state is s_2, the expected gain is $w_2(D)$. Now for any stationary policy, the corresponding one-step-removed policies are the same as the original, and so, in particular, in this case (2) implies that

$$w_1(D) = 10 + .9(.8w_1(D) + .2w_2(D))$$

$$w_2(D) = 4 + .9(.25w_1(D) + .75w_2(D)).$$

Solving this system of equations for $w_1(D)$ and $w_2(D)$, we have

$$w_1(D) \approx 78.61$$

$$w_2(D) \approx 66.73.$$

PROBLEM SET 3.2

1. Consider the Markov decision process of Table 2.1 of Section 2.2 as having an infinite horizon.

   a) How many distinct general stationary policies does the process have?

   b) Compute the values $w_1(D)$ and $w_2(D)$ of the stationary policy D defined by $D(1) = 1$ and $D(2) = 2$, using $\alpha = .9$.

   c) Compute the values of the stationary policy of the example of this section using $\alpha = 1/2$.

2. Consider the Markov decision process of Problem 3 of Section 3.1 to have an infinite horizon with discount factor α. Let D_1 be the stationary policy defined by $D_1(1) = 1$ and D_2 the stationary policy defined by $D_2(1) = 2$.

   a) Compute $w_1(D_1)$, $w_1(D_2)$, and $w_2(D_1) = w_2(D_2)$ using the definition (1).

   b) Use (2) to check your answers.

   c) Determine α_0 such that $w_1(D_1) > w_1(D_2)$ for $\alpha < \alpha_0$ and $w_1(D_1) < w_1(D_2)$ for $\alpha > \alpha_0$.

3. Consider the Markov decision process of Problem 4 of Section 3.1 to have an infinite horizon with discount factor α. Let D be the stationary policy defined by $D(1) = 2$.

   a) Determine the probabilities $p(n,j)$ for the policy D with initial state s_1.

   b) Using these probabilities compute $w_1(D)$ using the expression (1).

   c) Check your answer using (2).

4. (Alternate proof of the Lemma.) Let D, $\bar{D}_j$, and $k*$ be as defined in the Lemma, and $p(n,j)$ and $D(n,j)$ as defined in the definition of $w_1(D)$. Let $\bar{p}_j(n, \ell)$ be the probability that the process will be in state s_ℓ after n transitions if the process starts in state s_j and policy $\bar{D}_j$ is used.

   a) Show that

   $$w_j(\bar{D}_j) = q_j^{D(1,j)} + \sum_{n=1}^{\infty} \alpha^n \sum_{\ell=1}^{N} \bar{p}_j(n,\ell) \, q_\ell^{D(n+1),\ell}$$

   and

   $$p(n+1,\ell) = \sum_{j=1}^{N} p_{ij}^{k*} \bar{p}_j(n,\ell).$$

   b) Now using these two equations and the series expression for $w_i(D)$, start with the sum

$$\alpha \sum_{j=1}^{N} p_{ij}^{k} \; w_{j}(\overline{D}_{j})$$

and verify Equation (2).

3.3 The Stationary Policy Theorem

In Section 2.1, when working with a Markov decision process with a finite horizon, in order to define the quantity $v_i(n)$ we considered the value of each policy with initial state s_i and n transitions remaining and set $v_i(n)$ equal to the maximum. That would suggest that for a process with an infinite horizon and discount factor $\alpha < 1$, we consider the set

$\{w_i(D) \mid D$ a policy with initial state $s_i\}$.

However, as already observed in the last section, this set is no longer finite; nevertheless the quantity $v_i(n)$ still does have a very useful extension.

We proved in Section 3.2 that for any i and any policy D with initial state s_i,

$|w_i(D)| \leq Q/(1 - \alpha)$,

where Q is chosen so that $|q_{ij}^{k}| \leq Q$ for all i and k. Thus the set of $w_i(D)$'s is bounded, and so, for a Markov decision process with infinite horizon and discount factor α, $0 < \alpha < 1$, we can define

$v_i = 1\,u\,b\;\{w_i(D) \mid D$ a policy with initial state $s_i\}$

where l u b is the standard abbreviation for "least upper bound."

Note that since each v_i is defined as the least upper bound over an infinite set, it is no longer obvious that for each i there is an optimal policy, that is, a policy D with initial state s_i such that $w_i(D) = v_i$. But in fact there is always an optimal policy; in this section we will demonstrate the existence of a stationary policy D such that $w_i(D) = v_i$ for each i.

To prepare for this we first need an extension of the recursive relationship (1) of Section 3.1:

$$(1) \quad v_i(n) = \underset{1 \leq k \leq A_i}{\text{Max}} \{q_i^{k} + \alpha \sum_{j=1}^{N} p_{ij}^{k} \, v_j(n-1)\}.$$

If we knew that $v_i(n) \to v_i$ as $n \to \infty$, taking limits on both sides of (1) would yield

$$v_i = \underset{1 \le k \le A_i}{\text{Max}} \{q_i^k + \alpha \sum_{j=1}^{N} p_{ij}^k v_j\}.$$

This relationship is indeed valid, as we now prove.

THEOREM 1. For a Markov decision process with infinite horizon and discount factor α, $0 < \alpha < 1$, define

$$u_i = \underset{1 \le k \le A_i}{\text{Max}} \{q_i^k + \alpha \sum_{j=1}^{N} p_{ij}^k v_j\}, \quad 1 \le i \le N.$$

Then, for each i,

$$u_i = v_i.$$

PROOF. Fix i, $1 \le i \le N$.

First we show that $v_i \le u_i$: Let D be any policy with initial state s_i. Suppose D prescribes alternative k* on the first transition out of s_i; and denote by $\bar{D}_j$ the associated one-step-removed policies. Then, using the Lemma of the last section,

$$w_i(D) = q_i^{k*} + \alpha \sum_{j=1}^{N} p_{ij}^{k*} w_j(\bar{D}_j)$$

$$\le q_i^{k*} + \alpha \sum_{j=1}^{N} p_{ij}^{k*} v_j$$

$$\le \underset{1 \le k \le A_i}{\text{Max}} \{q_i^k + \alpha \sum_{j=1}^{N} p_{ij}^k v_j\} = u_i.$$

Therefore u_i is an upper bound for the set

$$\{w_i(D) \mid D \text{ a policy with initial state } s_i\}$$

and so

$$v_i = l u b \{w_i(D)\} \le u_i.$$

Second, we show that $u_i \le v_i$: Select alternative k_i such that

$$u_i = \underset{1 \le k \le A_i}{\text{Max}} \{q_i^k + \alpha \sum_{j=1}^{N} p_{ij}^k v_j\}$$

$$= q_i^{k_i} + \alpha \sum_{j=1}^{N} p_{ij}^{k_i} v_j.$$

Now take any $\varepsilon > 0$. For each j, choose a policy D_j^* with initial state s_j such that $v_j - \varepsilon < w_j(D_j^*)$. Define a policy D with initial state s_i as follows: use alternative k_i out of state s_i, and then, for each j, if the system moves to state s_j on the first transition, use policy D_j^* thereafter. Then from the Lemma of the last section we have

$$w_i(D) = q_i^{k_i} + \alpha \sum_{j=1}^{N} p_{ij}^{k_i} w_j(D_j^*)$$

and so

$$u_i = q_i^{k_i} + \alpha \sum_{j=1}^{N} p_{ij}^{k_i} v_j$$

$$\leq q_i^{k_i} + \alpha \sum_{j=1}^{N} p_{ij}^{k_i} (w_j(D_j^*) + \varepsilon)$$

$$= q_i^{k_i} + \alpha \sum_{j=1}^{N} p_{ij}^{k_i} w_j(D_j^*) + \alpha \varepsilon \sum_{j=1}^{N} p_{ij}^{k_i}$$

$$= w_i(D) + \alpha \varepsilon$$

$$< v_i + \varepsilon.$$

Since ε is arbitrary, $u_i \leq v_i$. ###

The relationship (2) suggests that for each state s_i there is a distinguished alternative, namely that alternative at which the

$$\underset{1 \leq k \leq A_i}{\text{Max}} \; \{q_i^k + \alpha \sum_{j=1}^{N} p_{ij}^k v_j\}$$

is attained; and in fact the stationary policy defined by these distinguished alternatives is precisely the optimal policy we seek.

THEOREM 2. (Stationary Policy Theorem.) Given a Markov decision process with infinite horizon and discount factor α, $0 < \alpha < 1$, choose, for each i, an alternative k_i such that

$$\underset{1 \leq k \leq A_i}{\text{Max}} \; \{q_i^k + \alpha \sum_{j=1}^{N} p_{ij}^k v_j\} = q_i^{k_i} + \alpha \sum_{j=1}^{N} p_{ij}^{k_i} v_j.$$

Define the stationary policy D by $D(i) = k_i$ (that is, use alternative k_i when in state s_i). Then for each i,

$$w_i(D) = v_i.$$

PROOF. (Note: We use M^t to denote the transpose of a matrix M.)

Define

$$P = \begin{bmatrix} p_{ij}^{k_i} \end{bmatrix}, \quad 1 \le i, \ j \le N,$$

the $N \times N$ matrix with i^{th} row consisting of the transition probabilities of alternative k_i out of state s_i.

Define

$$q = \begin{bmatrix} q_1^{k_1}, \ q_2^{k_2}, \ \ldots, \ q_N^{k_N} \end{bmatrix}^t,$$

the column vector made up of the expected rewards associated with the alternatives k_i.

Define

$$v = [v_1, \ v_2, \ \ldots, \ v_N]^t.$$

Now Theorem 1 states that for each i,

$$v_i = q_i^{k_i} + \alpha \sum_{j=1}^{N} p_{ij}^{k_i} v_j.$$

These equalities can be expressed in terms of the matrix and vectors we have defined, namely,

$$v = q + \alpha P v.$$

Thus

$$v - \alpha P v = q$$

and so

$$(I - \alpha P) v = q$$

where I is the $N \times N$ identity matrix.

Similarly, define $w_D = [w_1(D), \ w_2(D), \ \ldots, \ w_N(D)]^t$, the column vector consisting of the values associated with the stationary policy D. Now each one-step-removed policy $\bar{D}_j$ is the same as policy D, and so the Lemma of the last section states that for each i,

$$w_i(D) = q_i^{k_i} + \alpha \sum_{j=1}^{N} p_{ij}^{k_i} w_j(D).$$

In matrix terms, these equations read

$$w_D = q + \alpha P w_D.$$

Hence

$$w_D - \alpha P w_D = q$$

and so

$$(I - \alpha P) w_D = q.$$

We have proved that $(I - \alpha P)w_D = q = (I - \alpha P)v$. We want to show that $w_i(D) = v_i$ for each i, that is, that $w_D = v$. Now if we knew that the matrix $I - \alpha P$ were invertible we would be done, as we would have

$$w_D = (I - \alpha P)^{-1}q = v.$$

Thus the following lemma completes the proof of the theorem.

LEMMA. The matrix $I - \alpha P$ as defined above is invertible.

PROOF. By adopting the stationary policy D we have fixed the alternative to use for each state. Thus the progression of the process from state to state is as in that of a Markov chain with transition matrix P. The ij^{th} entry then of the matrix P^n is the probability of the system moving from state s_i to state s_j in n transitions (see Problem 6 of Section 1.1), and so is a nonnegative number bounded by one. Now for each n define an $N \times N$ matrix

$$S_n = I + \alpha P + \alpha^2 P^2 + \ldots + \alpha^n P^n.$$

Each entry of S_n is a sum of $(n + 1)$ nonnegative terms, and the sum is bounded by

$$1 + \alpha + \alpha^2 + \ldots + \alpha^n \leq \frac{1}{1-\alpha}.$$

Hence each entry of S_n is the $(n + 1)$ - partial sum of a convergent series, and so converges as $n \to \infty$. Denote the matrix of limiting values by S, that is,

$$S = \lim_{n \to \infty} S_n.$$

Now

$$S_n(I - \alpha P) = (I + \alpha P + \alpha^2 P^2 + \ldots + \alpha^n P^n)(I - \alpha P)$$

$$= I - \alpha^{n+1}P^{n+1}$$

and each entry of $\alpha^{n+1}P^{n+1}$ converges to zero as $n \to \infty$. Therefore

$$S(I - \alpha P) = (\lim_{n \to \infty} S_n)(I - \alpha P)$$

$$= \lim_{n \to \infty} S_n (I - \alpha P) \quad \text{(see Problem 4)}$$

$$= \lim_{n \to \infty} (I - \alpha^{n+1}P^{n+1})$$

$$= I.$$

Therefore $I - \alpha P$ is invertible, and $(I - \alpha P)^{-1} = S$. ###

It might seem that our work for processes with an infinite horizon should be complete. To determine a stationary policy that is optimal, no matter what the initial state is, we simply choose, for each i, an alternative k_i at which

$$\underset{1 \leq k \leq A_i}{\text{Max}} \{q_i^k + \alpha \sum_{j=1}^{N} p_{ij}^k v_j\}$$

is attained. However in order to proceed this way we would need to know the values of the v_j's, and we have not yet developed a practical method for computing them. The definition of each is in terms of the least upper bound of an intractable infinite set, and the expression (2) relates the v_j's only to each other. This expression (2) though can serve as the foundation for effective solution techniques, as we shall see in the next chapter.

The Stationary Policy Theorem does suggest one method of determining an optimal policy. From the theorem we know that there is a stationary policy that is optimal for any initial state, and there are only a finite number of stationary policies. Thus we could evaluate each policy; a policy delivering the maximum value for each starting state would be an optimal stationary policy. We demonstrate this approach in the following example; however for the general case the technique is quite impractical. There can be many stationary policies (in fact, their number is given by the product of the A_i's), and the computation of policy values involves the solving of a system of N equations and N unknowns (as in Example 1 of Section 3.2).

EXAMPLE 1. Consider the Markov decision process of Table 2.1 of Section 2.2 as having an infinite horizon and discount factor .9. There are six stationary policies; their values rounded off to two decimal places are given in Table 3.3.

TABLE 3.3

Policy D		Values	
D(1)	D(2)	$w_1(D)$	$w_2(D)$
1	1	77.74	66.13
1	2	80.50	69.50
1	3	78.92	67.57
2	1	78.61	66.73
2	2	80.27	69.32
2	3	79.46	68.04

From either the $w_1(D)$ column of values or the $w_2(D)$ column, we see that the stationary policy D defined by $D(1) = 1$, $D(2) = 2$ is the optimal policy. (Compare this result with the data for the discounted finite horizon case given in Table 3.2 of Section 3.1.) The optimal values are $v_1 = w_1(D) = 80.5$ and $v_2 = w_2(D) = 69.5$.

PROBLEM SET 3.3

1. The data of Example 1 correspond to the process developed for the weekly newspaper model in Chapter 1. Assuming we accept this (rather unrealistically high) discount rate of .9 for the model, what advice would you give the publisher on the policy to use and the total value of the publication if he/she wishes to extend to a year-round operation?

2. Consider the Markov decision process of Problem 2 of Section 2.2 to have an infinite horizon and discount factor $\alpha = 1/2$.

    a) Calculate the values of each stationary policy. (Note: For each you can follow the example in Section 3.2, or you can use the equivalent matrix solution

    $$w_D = (I - \alpha P)^{-1} q).$$

    b) Determine the optimal policy and the values v_1 and v_2.

3. Determine the optimal policy and values for the Markov decision process of Problem 1 of Section 2.2, assuming the process has an infinite horizon and discount factor .75.

4. Suppose $B_n = [b_{ij}(n)]$, $B = [b_{ij}]$, and $C = [c_{ij}]$ are 2×2 matrices and that $b_{ij}(n) \to b_{ij}$ as $n \to \infty$, $1 \leq i$, $j \leq 2$.

    a) Show that each entry of the product $B_n C$ converges to the corresponding entry of $B C$, that is, show that

    $$\lim_{n \to \infty} B_n C = B C.$$

    b) What theorems on convergent sequences do you use in Part a?

5. Let $S = (I - \alpha P)^{-1}$ be as defined in the proof of Theorem 2.

    a) Show that all the entries of S are nonnegative, and that the entries on the main diagonal are all greater than or equal to one.

    b) Show that the sum of the entries in each row is $1/(1 - \alpha)$.

6. Let P be the transition matrix defined in the proof of Theorem 2. Show that the matrix $I - P$ is not invertible.

7. Are the following statements concerning Markov decision processes with infinite horizon true or false?

   a) Suppose D_1 and D_2 are stationary policies such that for some i^*, $w_{i^*}(D_1) > w_{i^*}(D_2)$. Then $w_i(D_1) \geq w_i(D_2)$ for all i (as in Table 3.3)?

   b) Suppose D^* is a stationary policy such that for some i^*, $w_{i^*}(D^*) \geq w_{i^*}(D)$ for any other stationary policy D. Then D^* is optimal?

   Hint: Consider the process defined by the following data.

State	Alt	q_i^k	p_{i1}^k	p_{i2}^k	p_{i3}^k
1	1	3	0	0	1
	2	0	0	0	1
2	1	3	0	0	1
	2	0	0	0	1
3	1	0	0	0	1

-46-

4. Solution Techniques

In this chapter we develop three methods for determining optimal policies and values for Markov decision processes with infinite horizon and future rewards discounted. The reason for this proliferation of solution techniques is that no one method is ideal for all problems.

The first algorithm we develop, the Policy Iteration Algorithm, has the very desirable property of convergence after a finite number of steps to an optimal policy. However, the calculations necessary at each iteration involve the solving of a system of N equations and N unknowns, where N is the number of states. Thus if the system has a large number of states, this algorithm may be difficult if not impossible to implement.

The next algorithm that we consider is the Value Iteration Algorithm. Here the calculations necessary at each iteration are very elementary, and so the algorithm may be used when N is large. But the algorithm converges to the optimal values of a process only in the limit. Thus at each iteration of the algorithm we may have only approximations to the optimal values, and so we may have to estimate when to terminate the calculations.

The third solution technique involves the use of linear programming. In Section 4.4 we assume that the reader is familiar with the theory and techniques of linear programming, and we show that the simplex algorithm can be used to determine optimal values and policies for an infinite horizon process. However, for a given process, the

size of the optimization problem generated depends on the
number of states and the total number of alternatives in
the process. Thus this technique is suitable only for processes of reasonable size.

The theory associated with these solution techniques
could be developed using matrix algebra. However critical
mappings associated with a Markov decision process are in
fact contraction mappings. Because of this and other properties of these mappings, the development of the theory can
be considerably simplified by using these ideas. Thus in
the first section of this chapter, we define these mappings
and discuss some of their relevant properties.

4.1 Contraction Mappings

Initially in the discussion of the optimization problem for a Markov decision process with infinite horizon, we
considered all possible policies. However from the Stationary Policy Theorem, it follows that in determining an
optimal policy, we need only deal with the set of stationary policies. In fact to each stationary policy D we can
associate the policy value vector

$$w_D = (w_1(D), \ldots, w_N(D))^t,$$

and we seek a D such that

$$w_D = v = (v_1, \ldots, v_N)^t \in R^N.$$

To develop techniques to determine such a policy, we will
first define and work with two special maps from R^N to R^N.
As we shall see, the points w_D and v will have important
roles to play with respect to these maps.

For a stationary policy D, define the map $T_D: R^N \to R^N$
by

$$T_D(x) = q + \alpha Px \text{ for } x \in R^N,$$

where

$$q = [q_1^{D(1)}, \ldots, q_N^{D(N)}]^t$$

and

$$P = [p_{ij}^{D(i)}]$$

are the expected reward vector and transition matrix, respectively, associated with D.

We will denote the ith component of $T_D(x)$ by $(T_Dx)(i)$.
We have, by definition, for $x = (x_1, \ldots, x_N)$,

$$(T_D x)(i) = q_i^{D(i)} + \alpha \sum_{j=1}^{N} p_{ij}^{D(i)} x_j.$$

This quantity represents the expected gain if the process is in state s_i, has one transition remaining, expected rewards discounted, terminal values given by the vector x, and policy D is used.

Continuing with this interpretation, $T_D(T_D(x))$, which we will denote by $T_D^2(x)$, is simply

$$T_D(T_D(x)) = T_D(q + \alpha D x)$$

$$= q + \alpha P(q + \alpha D x)$$

$$= q + \alpha P q + \alpha^2 P^2 x.$$

Since the ij^{th} entry of P^2 is the probability of moving from s_i to s_j in two transitions using policy D, the ith component of $T_D^2(x)$ is the total expected gain if the process is in state s_i, has two transitions remaining and terminal values x, and policy D is used. Similarly, the components of $T_D^n(x) = T_D(T_D^{n-1}(x))$ provide the total expected gains using policy D over n transitions and then terminating with values x.

The definition of the second map that we need is motivated by the key relationship that the vector v satisfies, Equation (2) of Section 3.3. We define $T_\alpha: R^N \to R^N$ as follows: For any $x = (x_1, \ldots, x_N) \in R^N$, define the ith component of $T_\alpha(x)$ by

$$(T_\alpha x)(i) = \underset{1 \le k \le A_i}{\text{Max}} \{q_i^k + \alpha \sum_{j=1}^{N} p_{ij}^k x_j\}.$$

An interpretation can also be given the map T_α. Suppose the process is in state s_i, has one transition remaining and terminal values x, with future rewards discounted. Then the value of an optimal policy, denoted in Chapter 2 by $v_i(1)$, is

$$v_i(1) = \underset{1 \le k \le A_i}{\text{Max}} \{q_i^k + \alpha \sum_{j} p_{ij}^k x_j\} = (T_\alpha x)(i).$$

Similarly, if two transitions remain, from (1) of Section 3.1,

$$v_i(2) = \max_{1 \le k \le A_i} \{q_i^k + \alpha \sum_j p_{ij}^k \, v_j(1)\}$$

$$= \max_{1 \le k \le A_i} \{q_i^k + \alpha \sum_j p_{ij}^k (T_\alpha x)(j)\}$$

$$= \text{the } i^{th} \text{ component of } T_\alpha^2(x) = T_\alpha(T_\alpha(x)).$$

By induction we have that, for any n, the vector of optimal values

$$(v_1(n), \ldots, v_N(n)) = T_\alpha^n(x).$$

The maps T_D and T_α have two very useful properties. The first is monotonicity.

LEMMA 1. Suppose $x = (x_1, \ldots, x_N)$ and $y = (y_1, \ldots, y_N)$ are in R^N, and that $x \le y$ (i.e., $x_i \le y_i$ for $1 \le i \le N$). Then

a) $T_\alpha(x) \le T_\alpha(y)$.

b) For any stationary policy D, $T_D(x) \le T_D(y)$.

PROOF. For any i, $1 \le i \le N$, and k, $1 \le k \le A_i$, $x \le y$ implies that

$$q_i^k + \alpha \sum_j p_{ij}^k \, x_j \le q_i^k + \alpha \sum_j p_{ij}^k \, y_j.$$

The Lemma now follows immediately from the definitions of the ith components of T_α and T_D. ###

To understand the consequences of the second and more interesting property of these maps, we first define the concept of a fixed point. A point $x \in R^N$ is a _fixed point_ of a map $T: R^N \to R^N$ if $T(x) = x$. The maps T_α and T_D have notable fixed points. Theorem 1 of Section 3.3 states that, for each i,

$$v_i = \max_{1 \le k \le A_i} \{q_i^k + \alpha \sum_j p_{ij}^k \, v_j\}$$

that is, $T_\alpha(v) = v$. And in the proof of the Stationary Policy Theorem, we noted that for a stationary policy D,

$$w_D = q + \alpha P \, w_D$$

where q and P are, respectively, the reward vector and transition matrix for D, and so $T_D(w_D) = w_D$.

In fact, these points are the unique fixed points of their respective maps, and furthermore, they are the limit points of certain sequences. These results are all part of the general theory of contraction mappings, which applies, since as we shall see, T_D and T_α are contraction maps. We now define the concept of a contraction map, and develop the properties that we will need later in the chapter.

First, for $x = (x_1, \ldots, x_N) \in R^N$, define

$$\|x\| = \max_{1 \le i \le N} \{|x_i|\}.$$

This defines a norm on R^N (see Problem 2). All we will need however are the following properties, the proofs of which are left to the reader (Problem 1).

LEMMA 2. For $x \in R^N$

a) $\|x\| = 0$ if and only if $x = 0 \in R^N$.

b) $\|-x\| = \|x\|$.

c) If $\{x(n)\} = \{(x_1(n), \ldots, x_N(n))\}$ is a sequence in R^N such that

$$\|x(n)\| \to 0 \text{ as } n \to \infty, \text{ then } x(n) \to 0 \in R^N$$

(i.e., $x_i(n) \to 0$ for $1 \le i \le N$).

A map $T: R^N \to R^N$ is said to be a <u>contraction mapping</u> if there is a constant β, $0 < \beta < 1$, such that for any x and y in R^N,

$$\|T(x) - T(y)\| \le \beta \|x - y\|.$$

LEMMA 3. For any stationary policy D, the associated map T_D is a contraction mapping (with constant α).

PROOF. Take any x and y in R^N. For any i, we have

$$(T_D x)(i) - (T_D y)(i) = q_i^{D(i)} + \alpha \sum_j p_{ij}^{D(i)} x_j$$

$$- (q_i^{D(i)} + \alpha \sum_j p_{ij}^{D(i)} y_j)$$

$$= \alpha \sum_j p_{ij}^{D(i)} (x_j - y_j)$$

$$\le \alpha \sum_j p_{ij}^{D(i)} \|x - y\|$$

$$= \alpha \|x - y\|.$$

Interchanging x and y, we have

$$(T_D y)(i) - (T_D x)(i) \leq \alpha \|y - x\| = \alpha \|x - y\|.$$

Thus

$$|(T_D x)(i) - (T_D y)(i)| \leq \alpha \|x - y\|.$$

Since i is arbitrary,

$$\|T_D(x) - T_D(y)\| = \underset{i}{\text{Max}} \{|(T_D x)(i) - (T_D y)(i)|\}$$

$$\leq \alpha \|x - y\|. \quad \#\#\#$$

LEMMA 4. T_α is a contraction mapping (with constant α).

PROOF. Take any x and y in R^N. Fix i. Choose k_i such that

$$q_i^{k_i} + \alpha \sum_j p_{ij}^{k_i} x_j = \underset{1 \leq k \leq A_i}{\text{Max}} \{q_i^k + \alpha \sum_j p_{ij}^k x_j\}.$$

Then

$$(T_\alpha x)(i) - (T_\alpha y)(i) = \underset{k}{\text{Max}} \{q_i^k + \alpha \sum_j p_{ij}^k x_j\}$$

$$- \underset{k}{\text{Max}} \{q_i^k + \alpha \sum_j p_{ij}^k y_j\}$$

$$= q_i^{k_i} + \alpha \sum_j p_{ij}^{k_i} x_j - \underset{k}{\text{Max}} \{q_i^k + \alpha \sum_j p_{ij}^k y_j\}$$

$$\leq q_i^{k_i} + \alpha \sum_j p_{ij}^{k_i} x_j - (q_i^{k_i} + \alpha \sum_j p_{ij}^{k_i} y_j)$$

$$= \alpha \sum_j p_{ij}^{k_i} (x_j - y_j)$$

$$\leq \alpha \sum_j p_{ij}^{k_i} \|x - y\| = \alpha \|x - y\|.$$

We can now interchange the x and y, and proceed as in Lemma 3. $\#\#\#$

Now, in the general theory, if T: $R^N \rightarrow R^N$ is a contraction map, then T has a fixed point, and this point is unique and is the limit of any sequence of the form $T^n(x)$, for fixed $x \in R^N$. Since we already know that T_α and T_D have fixed points, we prove here only the latter two claims. (The existence part of the Fixed Point Theorem is discussed in Problem 10.)

THEOREM 1. A contraction map can have at most one fixed point.

PROOF. Suppose $T: R^N \to R^N$ is a contraction map with constant β, and that $T(x) = x$ and $T(y) = y$. Then

$$\|x - y\| = \|T(x) - T(y)\| \leq \beta \|x - y\|.$$

Thus

$$(1 - \beta) \|x - y\| \leq 0,$$

and since $\beta < 1$, we must have

$$\|x - y\| = 0.$$

Hence $x = y$. ###

COROLLARY 1. The optimal value vector $v = (v_1, \ldots, v_N)^t$ is the unique fixed point of the map T_α; and for any stationary policy D, the vector $w_D = (w_1(D), \ldots, w_N(D))^t$ is the unique fixed point of the map T_D.

As an application of these results we give an alternate proof of the Stationary Policy Theorem of Section 3.3. Let D be the (optimal) stationary policy defined in that theorem. In the proof of the theorem we first showed that $T_D(v) = v$ and $T_D(w_D) = w_D$, and then appealed to the Lemma of that section, which states that the matrix $I - \alpha P$ is invertible, to conclude that $v = w_D$. But now equality of the v and w_D follows immediately from the above Corollary 1, and so we have a proof of the theorem that does not require the invertibility of the matrix $I - \alpha P$.

Before we state the next theorem, we need an extension to an arbitrary map the notation we have used for the maps T_α and T_D. Suppose $T: R^N \to R^N$. Then the composition $T \circ T: R^N \to R^N$ is well defined; denote this map by T^2. Similarly, by induction, define $T^n: R^N \to R^N$, $T^n = T \circ T^{n-1}$.

THEOREM 2. Suppose $T: R^N \to R^N$ is a contraction map with fixed point y. Then for any $x \in R^N$,

$$\lim_{n \to \infty} T^n(x) = y.$$

PROOF. Suppose β, $0 < \beta < 1$, is the contraction constant associated with T. Then for any $x \in R^N$,

$$\|T(x) - y\| = \|T(x) - T(y)\| \leq \beta \|x - y\|$$

and, by induction,

$$\|T^n(x) - y\| = \|T^n(x) - T(y)\| \leq \beta \|T^{n-1}(x) - y\|$$

$$\leq \beta^n \|x - y\|.$$

But

$$\beta^n \|x - y\| \to 0,$$

and so

$$\|T^n(x) - y\| \to 0, \text{ as } n \to \infty.$$

Hence $T^n(x) \to y$, using Lemma 2c.   ###

COROLLARY 2.   For any $x \in R^N$,

a) $\lim\limits_{n \to \infty} T_\alpha^n(x) = v.$

b)   For any stationary policy D,

$$\lim\limits_{n \to \infty} T_D^n(x) = w_D.$$

PROBLEM SET 4.1

1.   Prove Lemma 2.

2.   A norm on R^N is a function that assigns to each $x \in R^N$ a real number $\|x\|$ such that

a) $\|x\| \geq 0$, and $\|x\| = 0$ if and only if $x = 0$.
b) $\|x + y\| \leq \|x\| + \|y\|$.
c) $\|rx\| = |r| \, \|x\|$ for $r \in R$.

Show that the function defined in this section,

$$\|x\| = \underset{1 \leq i \leq N}{\text{Max}} \{x_i\},$$

is in fact a norm on R_N.

3.   Which of the following define a norm on R^N?

a) $\|x\|_a = |x_1|.$

b) $\|x\|_b = \begin{cases} 0, & x = 0 \\ 1, & x \neq 0. \end{cases}$

c) $\|x\|_c = |x_1| + \dots + |x_N|.$

d) $\|x\|_d = \underset{1 \leq i \leq N}{\text{Min}} \{|x_i|\}.$

4.   For $x = (x_1, \dots, x_N) \in R^N$, define

$$\|x\|_e = [x_1^2 + \dots + x_N^2]^{1/2},$$

the standard Euclidean norm on R^N. Let $\|x\|$ be the norm defined in this section.

a) Show that there is a constant r such that for any $x \in R^N$,

$$\|x\| \leq \|x\|_e \leq r \|x\|.$$

b) Show that for a sequence $\{x(n)\}$ in R^N, the following are equivalent:

    i) $\|x(n)\| \to 0$ as $n \to \infty$

    ii) $\|x(n)\|_e \to 0$ as $n \to \infty$

    iii) $x_i(n) \to 0$ as $n \to \infty$ for each i, $1 \leq i \leq N$.

5. Suppose $0 < \beta < 1$. Using the definitions of this section, determine which of the following are contraction maps from R^2 to R^2:

a) $T_a(x_1,x_2) = (x_1,\beta x_2)$.

b) $T_b(x_1,x_2) = (0,\beta x_2)$.

c) $T_c(x_1,x_2) = (\beta x_2-7,-\beta x_1+3)$.

6. Suppose $f: [0,1] \to [0,1]$ is continuous. Prove that f has a fixed point. Hint: consider the function $g(x) = f(x) - x$.

7. True or false: For a Markov decision process, the point $(x_1, \ldots, x_N) = (v_1, \ldots, v_N)$ is the only solution to the N equations

$$x_i = \underset{1 \leq k \leq A_i}{\text{Max}} \{q_i^k + \alpha \sum_j p_{ij}^k x_j\} \, , \, 1 \leq i \leq N.$$

8. For a stationary policy D, suppose that x is a point in R^N such that $T_D(x) \geq x$. Prove that $w_D \geq x$.

9. Given two stationary policies D_1 and D_2 define a stationary policy D by

$$D(i) = \begin{cases} D_1(i), & \text{if } w_i(D_1) \geq w_i(D_2) \\ D_2(i), & \text{if } w_i(D_2) > w_i(D_1). \end{cases}$$

Prove that $w_i(D) \geq \text{Max} \{w_i(D_1), w_i(D_2)\}$ for all i.

(Hint: Use Problem 8 with $x = (x_1, \ldots, x_N)$ determined by

$$x_i = \text{Max} \{w_i(D_1), w_i(D_2)\}.)$$

10. Suppose $T: R^N \to R^N$ is a contraction map with constant β. Take any $x \in R^N$.

a) Show that the sequence $\{T^n(x)\}$ is a Cauchy sequence. (Hint: show that for any m and n,

$$\|T^{n+m}(x) - T^n(x)\| \leq \beta^n \|T^m(x) - x\|$$

and then write

$$\|T^m(x) - x\| = \|T^m(x) - T^{m-1}(x) + T^{m-1}(x) - \ldots - x\|$$

and use Problem 2b.)

b) Since R^N is complete it follows that the sequence $\{T^n(x)\}$ must converge. Let $\lim T^n(x) = y$. Show that $T(y) = y$. (Hint: use the inequality

$$\|T^{n+1}(x) - T(y)\| \leq \beta \|T^n(x) - y\|$$

the fact that $T^n(x) - y \to 0$, and Problem 4b.)

4.2 Policy Iteration

The only method that we presently have available for determining optimal values and policies for a Markov decision process with an infinite horizon is to list all stationary policies, compute their values, and select a policy that delivers the maximum set of values. The Policy Iteration Algorithm, which we develop in this section, refines this process considerably. Starting with a stationary policy and its values, each iteration of the algorithm consists of two operations: first, the determination of a new policy; and second, the calculation of the values of this policy. Then with this new policy and values, the cycle is repeated. The algorithm, in outline form, follows.

POLICY ITERATION ALGORITHM

0. Preliminaries

a) Select an initial stationary policy D. For example, for each i, define D(i) such that

$$q_i^{D(i)} = \underset{1 \leq k \leq A_i}{\text{Max}} \{q_i^k\}.$$

b) Determine the policy values $w_D = (w_1(D), \ldots, w_N(D))$.

1. Computations

a) Select, for each i, an alternative k_i such that

$$q_i^{k_i} + \alpha \sum_j p_{ij}^{k_i} w_j(D) = \underset{1 \leq k \leq A_i}{\text{Max}} \{q_i^k + \alpha \sum_j p_{ij}^k w_j(D)\}.$$

Define policy D* by $D^*(i) = k_i$.

b) Compute the policy values $w_{D^*} = (w_1(D^*), \ldots, w_N(D^*))$.

2. Test and Termination

   a) If $w_D = w_{D*}$, stop, D is an optimal policy.

   b) Otherwise, return to Step 1, using for D the new policy D* (i.e., set D = D*).

As we shall prove, policy values improve at each iteration, and so once a policy is considered it cannot reappear in the process. Since there are only a finite number of stationary policies, the algorithm must terminate, and we shall show that at that point an optimal policy has been attained.

With convergence in a finite number of steps assured, the algorithm may seem ideal. However this is not quite the case, as the calculation of policy values, the major operation of each iteration, requires the solving of a system of N equations and N unknowns. If the number of states is small, and our computer memory space is adequate, this algorithm may be quite feasible. However if N is large, the value iteration algorithm of the next section would probably be more suitable.

Before we prove the convergence of the algorithm to an optimal policy, we give an example demonstrating the mechanics of the process.

EXAMPLE 1. Consider the Markov decision process of Table 2.1 of Section 2.2 as having an infinite horizon and discount factor α = .9. (The optimal policy and values for this process were found in Example 1 of Section 3.3 simply by listing all stationary policies and values.)

Step 0. Selecting for the initial policy that policy that delivers the best immediate gains, we define D by D(1) = 1, D(2) = 1. For this example we can read off policy values from Table 3.3 of Section 3.3. Here we have $w_1(D)$ = 77.74, $w_2(D)$ = 66.13.

Step 1. a) For each state we compare the quantities

$$q_i^k + \alpha \sum_j p_{ij}^k w_j(D), \quad 1 \leq k \leq A_i.$$

Thus, for i = 1, we compute

$$q_1^1 + \alpha (p_{11}^1 w_1(D) + p_{12}^1 w_2(D))$$

$$= 13 + .9(.5(77.74) + .5(66.13)) = 77.74$$

and

$$q_1^2 + \alpha \ (p_{11}^2 w_1(D) + p_{12}^2 w_2(D))$$

$$= 10 + .9(.8(77.74) + .2(66.13)) = 77.88.$$

The larger occurs at $k = 2$, and so $k_1 = 2$.

Similarly, for $i = 2$:

$$k \qquad q_i^k + \alpha \ \sum p_{ij}^k w_j(D)$$

$$1 \quad 4 + .9(.25(77.74)$$

$$+ .75(66.13)) = 66.13$$

$$2 \quad 2 + .9(.5(77.74)$$

$$+ .5(66.13)) = 66.74 \leftarrow$$

$$3 \quad -2 + .9((6/7)(77.74)$$

$$+ (1/7)(66.13)) = 66.47.$$

Hence $k_2 = 2$.

Define D^* by $D^*(1) = 2$, $D^*(2) = 2$.

b) From Table 3.3, $w_1(D^*) = 80.27$, $w_2(D^*) = 69.32$.

Step 2. $w(D^*) \neq w(D)$, and so we set $D(1) = 2$, $D(2) = 2$, and return to Step 1 using now the policy values 80.27 and 69.32 when determining the next policy. (Notice in fact that this new policy has larger values, so that we have moved, as promised, to a better policy.)

Step 1. a)

i	k	$q_i^k + \alpha \sum p_{ij}^k w_j(D)$
1	1	80.32 $\leftarrow$
	2	80.27
2	1	68.85
	2	69.32 $\leftarrow$
	3	68.84

Define D^* by $D^*(1) = 1$, $D^*(2) = 2$.

b) $w_1(D^*) = 80.50$, $w_2(D^*) = 69.50$.

Step 2. $w(D^*) > w(D)$.

Set $D(1) = 1$, $D(2) = 2$, and return to Step 1.

Step 1. a) i   k $q_i^k + \alpha \sum p_{ij}^k w_j(D)$

	1	1	80.50 ←
		2	80.47
	2	1	69.03
		2	69.50 ←
		3	69.04

Thus $D^* = D$.

Step 2. $w_{D^*} = w_D$, and so $D(1) = 1$, $D(2) = 2$ is an optimal stationary policy.

The convergence proof consists of two parts. First we show that at each iteration we move to a policy that is as least as good as the present policy; and then second, we show that if the policy values do not improve, we have an optimal policy. Our proofs rely heavily on the results of the last section.

THEOREM 1. Suppose that in Step 1a of the Policy Iteration Algorithm the new policy D^* is different from policy D. Then $w_{D^*} \geq w_D$.

PROOF. From Step 1 of the algorithm, using $D^*(i)$ for k_i, we have

$$q_i^{D^*(i)} + \alpha \sum p_{ij}^{D^*(i)} w_j(D) = \underset{k}{\text{Max}} \{q_i^k + \alpha \sum p_{ij}^k w_j(D)\}$$

and so, in particular,

$$q_i^{D^*(i)} + \alpha \sum p_{ij}^{D^*(i)} w_j(D) \geq q_i^{D(i)} + \alpha \sum p_{ij}^{D(i)} w_j(D).$$

Consider the map T_{D^*} and the image point $T_{D^*}(w_D)$. For any i,

$$(T_{D^*} w_D)(i) = q_i^{D^*(i)} + \alpha \sum p_{ij}^{D^*(i)} w_j(D)$$

$$\geq q_i^{D(i)} + \alpha \sum p_{ij}^{D(i)} w_j(D) = w_i(D).$$

Thus $T_{D^*}(w_D) \geq w_D$. Using Lemma 1b from Section 4.1, it follows that

$$T_{D^*}^2(w_D) \geq T_{D^*}(w_D) \geq w_D$$

and, by induction, for any n, that

$$T_{D^*}^n(w_D) \geq w_D.$$

Therefore, from Corollary 2b of Section 4.1,

$$w_{D^*} = \lim_{n \to \infty} T_{D^*}^n (w_D) \geq w_D. \quad \#\#\#$$

THEOREM 2. Suppose that in Step 2 of the Policy Iteration Algorithm, $w_{D^*} = w_D$. Then D is an optimal policy.

PROOF. If $w_{D^*} = w_D$, we have, from the definitions of the policy D* and the map T_α, for any i,

$$w_i(D) = w_i(D^*) = q_i^{D^*(i)} + \alpha \sum p_{ij}^{D^*(i)} w_j(D^*)$$

$$= q_i^{D^*(i)} + \alpha \sum p_{ij}^{D^*(i)} w_j(D)$$

$$= \underset{k}{\text{Max}} \{q_i^k + \alpha \sum p_{ij} w_j(D)\}$$

$$= (T_\alpha w_D)(i).$$

Thus $w_D = T_\alpha(w_D)$. From Corollary 1 of Section 4.1, w_D equals the optimal value vector v, and D is optimal. ###

COROLLARY 1. If, in Step 1a of the Policy Iteration Algorithm, D* = D, then D is an optimal policy.

COROLLARY 2. The Policy Iteration Algorithm converges in a finite number of iterations to an optimal policy.

PROOF. If, in Step 2, $w_{D^*} \neq w_D$, then D* $\neq$ D, and from Theorem 1, it follows that $w_{D^*} \geq w_D$. Since $w_{D^*} \neq w_D$, we must also have $w_i(D^*) > w_i(D)$ for at least one i. Thus once a policy is considered and replaced in the algorithm, it can never reappear. But there are only a finite number of stationary policies, and so we eventually reach an iteration where $w_{D^*} = w_D$ and Theorem 2 applies. ###

PROBLEM SET 4.2

1. Explain why in Step 1a of the algorithm, when k = D(i), the resulting quantity

$$q_i^k + \alpha \sum p_{ij}^k w_j(D)$$

is simply $w_i(D)$.

2. a) Determine optimal policies and values for the following process, using $\alpha = 1/2$:

State	Alt	q_i^k	p_{i1}^k	p_{i2}^k	p_{i3}^k
1	1	1	0	1	0
	2	0	0	0	1
2	1	0	0	1	0
3	1	1	0	0	1

b) Under what conditions could we have $w_D = w_{D*}$ in Step 2a of the algorithm but $D \neq D*$?

c) Show that, if in Step 1a of the algorithm, the $D*$ is uniquely determined, and $D* \neq D$, then $w_{D*} \neq w_D$.

3. Use the Policy Iteration Algorithm to determine optimal values and policy for the following process, with $\alpha = 1/2$:

State	Alt	q_i^k	p_{i1}^k	p_{i2}^k
1	1	11	1	0
	2	13	1/2	1/2
2	1	0	1	0
	2	3	7/15	8/15
	3	4	0	1

4. Determine optimal values and policies using policy iteration for the process

State	Alt	q_i^k	p_{i1}^k	p_{i2}^k
1	1	18	1/3	2/3
	2	11	3/4	1/4
2	1	-15	1/2	1/2
	2	-22	4/5	1/5

assuming:

a) $\alpha = 1/2$.

b) $\alpha = 2/3$.

c) $\alpha = 3/4$.

5. Determine how easily the computer system available for your use can invert matrices. Does a built-in inverse function exist; or would a subroutine be needed to calculate inverses? In either case, how large a matrix could the system effectively handle?

4.3 Value Iteration

In Chapter 2, the quantity $v_i(n)$ was defined to be the value of an optimal policy for a Markov decision process with initial state s_i and n transitions remaining, and in Section 3.1 this definition was modified to incorporate the presence of a discount factor. These values were easily computed iteratively using relationship (1) of Section 3.1.

It seems reasonable to expect that as n grows large, the quantity $v_i(n)$ should approximate v_i, the value of an optimal policy over the infinite horizon. In fact, this is the case; and moreover, we have already proven it in Section 4.1. For in the discussion following the definition of the map T_α, it was noted that $T_\alpha^n(x) = (v_1(n), \ldots, v_N(n))$ for any process with finite horizon, discount factor α, and terminal values x; and Corollary 2a of that section states that $T_\alpha^n(x) \to v$ as $n \to \infty$, for any $x \in R^N$. Thus we have, for $1 \leq i \leq N$,

$$\lim_{n \to \infty} v_i(n) = v_i.$$

This suggests another approach for determining optimal values. We can compute the quantities $v_i(n)$ iteratively, as was done in Chapter 2 and Section 3.1; and then use the $v_i(n)$'s to approximate the v_i's for n large. The algorithm is outlined below. (Since $T_\alpha^n(x) \to v$ for any terminal value vector x, we use terminal values of 0 when calculating the $v_i(n)$'s.)

VALUE ITERATION ALGORITHM

0. Preliminaries

Set $v_i(0) = 0$, $1 \leq i \leq N$, and $n = 1$.

1. Computations

Set, for each i

$$v_i(n) = \underset{1 \leq k \leq A_i}{\text{Max}} \{q_i^k + \alpha \sum_j p_{ij}^k v_j(n-1)\}.$$

2. Test and Termination

a) If $\underset{i}{\text{Max}} \{|v_i(n) - v_i(n-1)|\}$ is small, stop.

Use the values $v_i(n)$ to approximate the v_i's.

b) Otherwise, return to Step 1, setting $n = n + 1$ (i.e., compute the next set of $v_i(n)$'s).

Note the similarity between this algorithm and the Policy Iteration Algorithm. At each iteration we are concerned with quantities of the form

$$\underset{1 \leq k \leq A_i}{\text{Max}} \; \{q_i^k + \alpha \sum_j p_{ij}^k \, x_j\}.$$

In the Policy Iteration Algorithm, the x_j's are the values of the last policy, and the alternatives at which each maximum is attained define the next policy. In the Value Iteration Algorithm, the x_j's are the previous values calculated, and the values of each maximum define the next set of $v_i(n)$'s. However here the calculations are straightforward. Once the values have been determined, we move to the next step and calculate new values; no system of equations needs to be resolved. Thus the Value Iteration Algorithm is ideal for large problems.

We have already used the algorithm to compute the $v_i(n)$'s for small n. In particular Table 3.2 of Section 3.1 lists $v_1(n)$ and $v_2(n)$ for n up to 6 for the Markov decision process of Table 2.1 of Section 2.2, using $\alpha = .9$. Of course $v_1(6) = 41.23$ is not near $v_1 = 80.50$ (from Example 1 of Section 3.3), and similarly $v_2(6) = 30.23$ is not near $v_2 = 69.50$. But this is to be expected; larger values of n must be used in order to have the $v_i(n)$'s approximate the v_i's.

(Compare $\sum_{n=0}^{5} (.9)^n \approx 4.69$ with $\sum_{n=0}^{\infty} (.9)^n = 10$.)

Table 4.1 lists the same calculations for n = 10, 20, ..., 100.

TABLE 4.1

n	$v_1(n)$	$v_2(n)$
10	54.737	43.737
20	71.517	60.517
30	77.368	66.368
40	79.408	68.408
50	80.119	69.119
60	80.367	69.367
70	80.454	69.454
80	80.484	69.484
90	80.494	69.494
100	80.498	69.498

These data underscore an obvious question with the algorithm. Step 2a calls for termination of the calculations if the difference between successive $v_i(n)$'s is small. But what is small? One method of determining if we have stopped prematurely would be to consider the stationary policy D that delivered the last set of $v_i(n)$'s. We could calculate the values of this policy, and then determine if this policy is optimal using the test associated with the Policy Iteration Algorithm (i.e., check if, for each i, the

$$\underset{1 \leq k \leq A_i}{\text{Max}} \ \{q_i^k + \alpha \sum_j p_{ij}^k \ w_j(D)\}$$

is attained at $k = D(i)$). For example, in the above, for any $n \geq 2$, the stationary policy associated with each iteration is the optimal policy $D(1) = 1$, $D(2) = 2$. In fact we will show (Theorem 2) that for any infinite horizon process, there is an n_0 such that for any $n \geq n_0$, the alternatives at which each

$$\underset{k}{\text{Max}} \ \{q_i^k + \alpha \sum p_{ij}^k \ v_j(n-1)\}$$

is attained define an optimal stationary policy. But the theorem will not provide a practical means of computing the n_0; and moreover, implementation of this test procedure would require the determination of policy values and thus decrease considerably the computational simplicity of the algorithm.

There is a much more effective means of estimating the proximity of the data generated to the $v_i(n)$'s. Using successive values of the $v_i(n)$'s, upper and lower bounds on the v_i's can be easily calculated, and moreover, these bounds converge monotonically to the v_i's. We state and prove these results in Theorem 1. The following lemmas will be used in the proofs of this theorem and Theorem 2.

LEMMA 1. Define $e = (1, \ldots, 1)^t \ \epsilon \ R^N$. For any $x \ \epsilon \ R^N$ and $r \ \epsilon \ R$,

a) $T_\alpha(x + re) = T_\alpha(x) + \alpha re$.

b) For any stationary policy D, $T_D(x + re) = T_D(x) + \alpha re$.

PROOF. For any i, $1 \leq i \leq N$, and k, $1 \leq k \leq A_i$,

$$q_i^k + \alpha \sum_j p_{ij}^k (x_j + r) = q_i^k + \alpha \sum_j p_{ij}^k x_j + \alpha r.$$

The Lemma now follows immediately from the definitions of the ith components of T_α and T_D. ###

LEMMA 2. For any $x \in R^N$ and any r and s in R,

a) $x + re \leq T_\alpha(x) \leq x + se$ implies that

$$x + \frac{r}{1-\alpha} e \leq v \leq x + \frac{s}{1-\alpha} e.$$

b) For any stationary policy D,

$$x + re \leq T_D(x) \leq x + se \text{ implies that}$$

$$x + \frac{r}{1-\alpha} e \leq w_D \leq x + \frac{s}{1-\alpha} e.$$

PROOF. Assume that $T_\alpha(x) \leq x + se$. Then

$$T_\alpha^2(x) \leq T_\alpha(x + se) = T_\alpha(x) + \alpha se$$
$$\leq x + se + \alpha se$$
$$= x + s(1 + \alpha)e.$$

Continuing,

$$T_\alpha^3(x) = T_\alpha(T_\alpha^2(x))$$
$$\leq T_\alpha(x + s(1 + \alpha)e)$$
$$= T_\alpha(x) + s\alpha(1 + \alpha)e$$
$$\leq x + s(1 + \alpha + \alpha^2)e$$

and, by induction, for any n,

$$T_\alpha^n(x) \leq x + s(1 + \alpha + \dots + \alpha^{n-1})e$$
$$= x + (s \sum_{\lambda=0}^{n-1} \alpha^\lambda)e.$$

Letting $n \to \infty$, we have

$$v \leq x + \frac{s}{1-\alpha} e.$$

The other inequalities follow in the same fashion. ###

THEOREM 1. For each n, define

$$r_n = \underset{1 \leq i \leq N}{\text{Min}} \{v_i(n) - v_i(n - 1)\}$$

and

$$\overline{r}_n = \underset{1 \leq i \leq N}{\text{Max}} \{v_i(n) - v_i(n - 1)\}.$$

Then, for each i

$$v_i(n) + (\alpha r_n)/(1 - \alpha) \leq v_i \leq v_i(n) + (\alpha \overline{r}_n)/(1 - \alpha).$$

Moreover, these bounds are monotonic, that is,

$$v_i(n) + (\alpha r_n)/(1 - \alpha) \leq v_i(n+1) + (\alpha r_{n+1})/(1 - \alpha)$$

and

$$v_i(n + 1) + (\alpha \bar{r}_{n+1})/(1 - \alpha) \leq v_i(n) + (\alpha \bar{r}_n)/(1 - \alpha).$$

PROOF. Define $v(n) = (v_1(n), \ldots, v_N(n))^t \varepsilon\ R^N$. Then from the definition of r_n, we have

$$v(n - 1) + r_n e \leq v(n).$$

Thus

$$T_\alpha(v(n - 1) + r_n e) \leq T_\alpha(v(n)).$$

But, using Lemma 1a and the definition of T_α,

$$T_\alpha(v(n - 1) + r_n e) = T_\alpha(v(n - 1)) + \alpha r_n e$$

$$= v(n) + \alpha r_n e.$$

Hence

(1) $$v(n) + \alpha r_n e \leq T_\alpha(v(n)).$$

It follows immediately from Lemma 2a that

$$v(n) + [(\alpha r_n)/(1 - \alpha)]e \leq v,$$

that is,

$$v_i(n) + (\alpha r_n)/(1 - \alpha) \leq v_i$$

for each i, $1 \leq i \leq N$.

Similarly we can prove that, for each i,

$$v_i \leq v_i(n) + (\alpha \bar{r}_n)/(1 - \alpha). \quad \text{(Problem 2a)}$$

To show monotonicity, note that (1) above states that

$$v(n) + \alpha r_n e \leq T_\alpha(v(n)) = v(n + 1)$$

and so it follows from the definition of r_{n+1} that $\alpha r_n \leq r_{n+1}$. And, for each i, we have $v_i(n) + r_{n+1} \leq v_i(n + 1)$. Thus,

$$v_i(n + 1) + \frac{\alpha}{1-\alpha} r_{n+1} \geq v_i(n) + r_{n+1} + \frac{\alpha}{1-\alpha} r_{n+1}$$

$$= v_i(n) + \frac{1}{1-\alpha} r_{n+1}$$

$$\geq v_i(n) + \frac{\alpha}{1-\alpha} r_n.$$

Similarly for the upper bounds (Problem 2b). ###

These bounds for the v_i can be explained as follows. The quantity r_n is the present value of the minimum amount gained in the nth step of the process. Since gains earned in this n^{th} step are discounted by the factor α^{n-1}, the quantity $(r_n)/\alpha^{n-1}$ represents the minimal actual gain on n^{th} step. As the n increases, these actual gains should only increase; for example, the above inequality $\alpha r_n \leq r_{n+1}$ yields immediately the actual gain inequality $(r_n)/\alpha^{n-1} \leq (r_{n-1})/\alpha^n$. Thus $(r_n)\alpha^{n-1}$ is a lower bound on actual gains for each step beyond n, and so the sum

$$v_i(n) + \sum_{\lambda=n}^{\infty} [(r_n)/\alpha^{n-1}]\, \alpha^{\lambda}$$

$$= v_i(n) + [(r_n)/\alpha^{n-1}] \sum_{\lambda=n}^{\infty} \alpha^{\lambda}$$

$$= v_i(n) + (\alpha r_n)/(1 - \alpha)$$

should be a lower bound on the maximal accumulated gain for the process with discount factor α and initial state s_i, that is, on v_i.

These bounds are listed in Table 4.2 for the example referred to earlier in this section. The values are given for n = 2 and 3; the reason for stopping at such a small n is evident from the data. While our example is certainly elementary, the results are still striking.

TABLE 4.2

[Notation: $L_i(n) = v_i(n) + (\alpha r_n)/(1 - \alpha)$

$U_i(n) = v_i(n) + (\alpha \overline{r}_n)/(1 - \alpha)$]

n	$v_1(n)$	$L_1(n)$	$U_1(n)$	$v_2(n)$	$L_2(n)$	$U_2(n)$
2	20.65	71.50	89.50	9.65	60.50	78.50
3	26.64	80.50	80.50	15.64	69.50	69.50

Table 4.3 lists the bounds of the Markov decision process of Table 1.2 of Section 1.3, using a discount factor of $\alpha = .75$.

TABLE 4.3

n	$v_1(n)$	$L_1(n)$	$U_1(n)$
2	150.813	149.313	324.250
3	186.051	225.426	291.766
4	209.488	263.103	279.798
5	226.355	276.916	277.785
6	239.040	277.051	277.093
7	248.551	277.085	277.088
8	255.685	277.086	277.086

n	$v_2(n)$	$L_2(n)$	$U_2(n)$
2	78.000	76.500	251.438
3	91.125	130.500	196.840
4	108.997	162.613	179.307
5	126.140	176.701	177.569
6	138.810	176.822	176.863
7	148.323	176.857	176.859
8	155.456	176.857	176.857

n	$v_3(n)$	$L_3(n)$	$U_3(n)$
2	-24.500	-26.000	148.938
3	7.094	46.469	112.809
4	30.303	83.919	100.613
5	47.156	97.717	98.586
6	59.840	97.851	97.893
7	69.351	97.886	97.888
8	76.485	97.886	97.886

We conclude this section with a result of theoretical interest. When initially considering when to terminate calculations in the Value Iteration Algorithm, it was suggested that we could, for any n, consider the stationary policy associated with the nth iteration, that is, the policy defined by the alternatives at which each maximum value is attained, and determine if this policy is optimal for the infinite horizon process. This suggests an obvious question: for n large enough is any such policy optimal? We have seen in fact in the nondiscounted case that these policies need not become fixed with increasing n (Problems 8-10 of Section 2.2), and Problem 4 of this section provides a similar example for a discounted process. Nevertheless the answer to the question is yes.

THEOREM 2. Let the quantities $v_i(n)$ be defined by the Value Iteration Algorithm. Then there exists an n_0 such that for any $n \geq n_0$, if D* is a stationary policy satisfying, for each i, $1 \leq i \leq N$,

$$q_i^{D*(i)} + \alpha \sum p_{ij}^{D*(i)} v_j(n) = \underset{1 \leq k \leq A_i}{\text{Max}} \{q_i^k + \alpha \sum p_{ij}^k v_j(n)\}$$

then D* is an optimal policy for the infinite horizon process.

PROOF. Let

$$\varepsilon = \text{Min } \{\|v - w_D\| \,|\, D \text{ a non-optimal stationary policy}\}$$

Then $\varepsilon > 0$; choose $\delta > 0$ so that $2\delta/(1 - \alpha) < \varepsilon$. Since

$$v(n) = (v_1(n), \ldots, v_N(n)) \to (v_1, \ldots, v_N) \text{ as } n \to \infty,$$

there is an n_0 such that, for any $n \geq n_0$, $\|v(n) - v\| < \delta$, or equivalently,

$$v(n) < v + \delta e \quad \text{and} \quad v < v(n) + \delta e,$$

where $e = (1, \ldots, 1)^t \in R^N$, as before.

Now suppose policy D* satisfies the conditions of the theorem for some $n \geq n_0$. Then for each i we have

$$(T_{D*}v(n))(i) = q_i^{D*(i)} + \alpha \sum p_{ij}^{D*(i)} v_j(n)$$

$$= \underset{k}{\text{Max}} \{q_i^k + \alpha \sum p_{ij}^k v_j(n)\}$$

$$= (T_\alpha v(n))(i)$$

and so

$$T_{D*}(v(n)) = T_\alpha(v(n)).$$

Thus, using Lemma 1,

$$T_{D*}(v(n)) = T_\alpha(v(n))$$

$$\leq T_\alpha(v + \delta e)$$

$$= T_\alpha(v) + \alpha\delta e$$

$$= v + \alpha\delta e$$

$$\leq v(n) + \delta(1 + \alpha)e.$$

Using Lemma 2b it follows that

$$w_{D*} \leq v(n) + \frac{1+\alpha}{1-\alpha}\delta e$$

$$< v + \delta e + \frac{1+\alpha}{1-\alpha}\delta e$$

$$= v + \frac{2\delta}{1-\alpha}e.$$

Similarly we can show that

$$w_{D*} > v - \frac{2\delta}{1-\alpha}e.$$

Thus

$$\|w_{D*} - v\| < \frac{2\delta}{1-\alpha} < \varepsilon,$$

and so D* must be optimal. ###

PROBLEM SET 4.3

1. Prove Part b of Lemma 2.

2. Complete the proof of Theorem 1. In particular show that

a) $v_i \leq v_i(n) + (\alpha \overline{r}_n)/(1 - \alpha)$.

b) $v_i(n + 1) + (\alpha \overline{r}_{n+1})/(1 - \alpha) \leq v_i(n) + (\alpha \overline{r}_n)/(1 - \alpha)$.

3. Prove that the bounds of Theorem 1 do in fact converge to v_i, that is, show that

$$v_i(n) + (\alpha \underline{r}_n)/(1 - \alpha) \to v_i \text{ and}$$

$$v_i(n) + (\alpha \overline{r}_n)/(1 - \alpha) \to v_i \text{ as } n \to \infty.$$

4. For the Markov decision process given by the following data, with discount factor $\alpha = 1/2$,

State	Alt	q_i^k	p_{i1}^k	p_{i2}^k	p_{i3}^k	p_{i4}^k	$v_i(0)$
1	1	1	0	1	0	0	0
	2	0	0	0	1	0	
2	1	3	0	1	0	0	0
3	1	6	0	0	0	1	0
4	1	0	0	0	1	0	0

a) Show that if the horizon is finite with n transitions remaining, and the system is in state 1, the total expected reward if alternative 1 is used on the first transition is

$$4 - 3(1/2)^{n-1}.$$

b) Similarly show that if alternative 2 is used, the total expected reward is

$$4 - (1/4)^{k-1}$$

where $k = n/2$ if n is even and $k = (n - 1)/2$ if n is odd.

c) Show that

$$a_1(n) = \begin{cases} 1, & n \text{ odd} \\ 2, & n \text{ even.} \end{cases}$$

d) Suppose now the horizon is infinite. Find all optimal stationary policies.

5. Assuming a discount factor of $\alpha = 1/2$ for the Markov decision process

State	Alt	q_i^k	p_{i1}^k	p_{i2}^k	$v_i(0)$
1	1	1	1	0	0
	2	0	0	1	
2	1	2	0	1	0

a) Show that $a_1(n) = 1$ for all n. Conclusion: the stationary policy D, $D(1) = 1$, is an optimal policy for the infinite horizon process.

b) Show that the stationary policy $D\star$, $D\star(1) = 2$, is also an optimal policy for the infinite horizon process.

6. The first five iterations of the Value Iteration Algorithm applied to the process of Problem 3 of Section 4.2 give the following data:

n	1	2	3	4	5
$v_1(n)$	13	17.5	19.75	20.875	21.4375
$v_2(n)$	4	7.1	8.9767	10.0021	10.5381

Determine the bounds $L_i(n)$ and $U_i(n)$ for $i = 1, 2$; $n = 2, 3, 4, 5$. Using these bounds estimate v_1 and v_2.

7. Use the bounds generated by the Value Iteration Algorithm to determine optimal values for the processes of Problem 4 of Section 4.2.

8. Problem 12 of Section 2.2 called for the programming of the Value Iteration Algorithm. Now modify your program to incorporate a discount factor and to determine the error bounds for the optimal values for the infinite horizon process.

9. Use your program of Problem 8 to determine for how much the stove dealer of Example 3 of Section 1.3 should sell the business, assuming that it is a year-round operation, two stoves are presently in stock, and the discount factor is .98.

4.4 Linear Programming

Theorem 1 of Section 3.2 states that, for each i, the optimal values for the discounted infinite horizon Markov decision process satisfy the relationship

$$v_i = (T_\alpha v)(i) = \underset{1 \leq k \leq A_i}{Max} \{q_i^k + \alpha \sum_j p_{ij}^k v_j\}$$

and so we have

$$v_i \geq q_i^k + \alpha \sum_j p_{ij}^k v_j$$

for each k, $1 \leq k \leq A_i$. In fact, the v_i's are the smallest numbers that satisfy these ΣA_i inequalities. For suppose x $= (x_1, \ldots, x_N) \, \varepsilon \, R^N$ is a solution to the system

$$x_i \geq q_i^k + \alpha \sum_j p_{ij}^k x_j, \quad 1 \leq i \leq N, \, 1 \leq k \leq A_1.$$

Then, for each i,

$$x_i \geq \underset{1 \leq k \leq A_i}{Max} \{q_i^k + \alpha \sum_j p_{ij}^k x_j\} = (T_\alpha x)(i)$$

and so $x \geq T_\alpha(x)$; it follows immediately from Lemma 2a of Section 4.3 that $x \geq v$. This suggests that the problem of determining the v_i's can be translated into an optimization problem involving this family of inequalities.

THEOREM. The unique solution point to the linear programming problem of

Minimizing

$$x_1 + x_2 + \ldots + x_N$$

subject to

(1) $\qquad x_i \geq q_i^k + \alpha \sum_j p_{ij}^k x_j, \, 1 \leq i \leq N, \, 1 \leq k \leq A_i$

is the point $x = v = (v_1, \ldots, v_N)$.

PROOF. First note that $v = (v_1, \ldots, v_N)$ satisfies the constraints of (1). Now suppose $x = (x_1, \ldots, x_N)$ is an optimal solution point to the linear programming problem. Since x satisfies (1), from the above we have $x \geq v$, and so $x_i \geq v_i$ for each i. If $x \neq v$, then for some i*, $x_{i*} > v_{i*}$, and so $\Sigma x_i > \Sigma v_i$, a contradiction to the optimality of x. ###

Thus the simplex algorithm can be used to determine the optimal values v_i for a discounted, infinite horizon, Markov decision process. For large problems this technique may not be suitable, though, for the associated linear programming problem has N original variables and ΣA_i inequalities. However if a programmed version of the algorithm is available and the problem at hand is only of moderate size, we have a viable solution technique.

One minor note in passing. The above linear programming problem does not have the usual nonnegativity restrictions on the variables involved. This may be a problem with using the solution technique if the only available version of the simplex algorithm assumes nonnegativity. Of course if we know beforehand that the solution point v must have all nonnegative coordinates (e.g., if all the $q_i^k \geq 0$), then since all x satisfying (1) are greater than or equal to v, we can add the N restrictions $x_i \geq 0$ to the problem with no loss of generality.

However in the case when one or some of the v_i's might be negative, modifications would be necessary if nonnegative variables are required. We could replace each arbitrary variable x_i with the difference of two nonnegative variables, a standard technique; or we could shift the variables, substituting say, for each i, $x_{i*} = x_i + b$, where b is estimated so that $v_i + b \geq 0$ for all i. Perhaps the simplest method though, in this situation, is to add a constant, say c, to all the q_i^k's, c chosen so that the resulting modified process would have nonnegative optimal values. For example, if

$$c = \text{Max } \{-q_i^k \mid q_i^k < 0\}$$

then $q_i^k + c \geq 0$ for all i and k. The addition of the constant c to all the gains will increase each optimal value of the resulting infinite horizon process by the amount $c/(1 - \alpha)$, because we are increasing our actual gain at each step by c, and the accumulated present value of this is

$$c + c\alpha + c\alpha^2 + \ldots = \sum_{n=0}^{\infty} c\alpha^n = \frac{c}{1-\alpha} .$$

Thus simply subtracting this quantity from the optimal values of the modified process will give the optimal values for the original process.

EXAMPLE 1. The linear programming problem associated with the Markov decision process of Table 2.1 in Section 2.2, with $\alpha = .9$ (our standard example), is to

Minimize
$$x_1 + x_2$$

subject to

$$x_1 - .9((1/2)x_1 + (1/2)x_2) \geq 13$$

$$x_1 - .9((4/5)x_1 + (1/5)x_2) \geq 10$$

(2) $$x_2 - .9((1/4)x_1 + (3/4)x_2) \geq 4$$

$$x_2 - .9((1/2)x_1 + (1/2)x_2) \geq 2$$

$$x_2 - .9((6/7)x_1 + (1/7)x_2) \geq -2.$$

Combining fractions, the constraints reduce to

$$\frac{11}{20} x_1 - \frac{9}{20} x_2 \geq 13$$

$$\frac{7}{25} x_1 - \frac{9}{50} x_2 \geq 10$$

(3) $$-\frac{9}{40} x_1 + \frac{13}{40} x_2 \geq 4$$

$$-\frac{9}{20} x_1 + \frac{11}{20} x_2 \geq 2$$

$$-\frac{27}{35} x_1 + \frac{61}{70} x_2 \geq -2.$$

Since any stationary policy that does not use Alternative 3 out of state s_2 must have positive values, we know that v_1 and $v_2 > 0$, and so the restrictions $x_1 \geq 0$ and $x_2 \geq 0$ can be added to the problem without loss of generality. The resulting dual linear programming problem is to

Maximize
$$13y_1 + 10y_2 + 4y_3 + 2y_4 - 2y_5$$

subject to

$$(4) \quad \frac{11}{20}\, y_1 + \frac{7}{25}\, y_2 - \frac{9}{40}\, y_3 - \frac{9}{20}\, y_4 - \frac{27}{35}\, y_5 \leq 1$$

$$-\frac{9}{20}\, y_1 - \frac{9}{50}\, y_2 + \frac{13}{40}\, y_3 + \frac{11}{20}\, y_4 + \frac{61}{70}\, y_5 \leq 1$$

$$y_1, \ \ldots, \ y_5 \geq 0.$$

This dual problem may be more easily resolved. It has only two constraints (in general, there would be only N), and the introduction of slack variables into these two "$\leq$" inequalities eliminates the need for artificial variables.

In using the simplex algorithm to resolve either problem, a solution to each would be generated; here it would be determined that $(x_1, x_2) = (80.5, 69.5)$ is a solution to the original problem and $(y_1, \ldots, y_5) = (10, 0, 0, 10, 0)$ is a solution to the dual. Thus we would have immediately the optimal values $v_1 = 80.5$ and $v_2 = 69.5$.

Suppose we also need to determine an optimal stationary policy. We know from the Stationary Policy Theorem of Section 3.3 that the policy defined by using, for each i, an alternative that delivers the

$$\underset{k}{\text{Max}} \ \{q_i^k + \alpha \sum_j p_{ij}^k\, v_j\}$$

is optimal. Now for each i this maximum is simply v_i, so all we need determine is that k, $1 \leq k \leq A_i$, for which

$$v_i = q_i^k + \alpha \sum_j p_{ij}^k\, v_j.$$

Notice that if we let $x = v$, these quantities

$$q_i^k + \alpha \sum_j p_{ij}^k\, v_j$$

are all listed on the right side of (1), and that, in fact, what we seek, for each i, is that inequality in (1) among the A_i associated with state s_i that is an equality when $x = v$.

In the example, the first two inequalities in (2), and so in (3), are generated from the two alternatives out of state s_1. Letting $x_1 = 80.5$ and $x_2 = 69.5$ in (3), they yield

$$\frac{11}{20}\,(80.5) - \frac{9}{20}\,(69.5) = 13 = 13$$

$$\frac{7}{25}\,(80.5) - \frac{9}{50}\,(69.5) = 10.03 > 10.$$

Thus an optimal policy D must have $D(1) = 1$.

Similarly the last three inequalities in (3), with $(x_1, x_2) = (80.5, 69.5)$, yield

$$-\frac{9}{40}(80.5) + \frac{13}{40}(69.5) = 4.475 > 4$$

$$-\frac{9}{20}(80.5) + \frac{11}{20}(69.5) = 2 = 2$$

$$-\frac{27}{35}(80.5) + \frac{61}{70}(69.5) = -1.5357 > -2$$

and so $D(1) = 1$, $D(2) = 2$ defines the optimal stationary policy.

We can determine an optimal stationary policy, however, without performing these calculations. The Complementary Slackness Theorem of Linear Programming (see for example Theorem 4, p. 136 of [5] or Problem 7, p. 125, of [20]) states that a necessary and sufficient condition for a pair of feasible solutions to the primal and dual problems, respectively, to be optimal is that all possible products of the slack in a constraint, when evaluated at the feasible solution, times the value of the dual variable corresponding to constraint, must be zero. In our example this means that in order for (x_1, x_2) to be an optimal solution to the linear programming problem (3) and $(y_1, \ldots, y_5)$ to be an optimal solution to the dual problem (4), we must have

$$y_1 \left(\frac{11}{20} x_1 - \frac{9}{20} x_2 - 13 \right) = 0$$

$$y_2 \left(\frac{7}{25} x_1 - \frac{9}{50} x_2 - 10 \right) = 0$$

$$y_3 \left(-\frac{9}{40} x_1 + \frac{13}{40} x_2 - 4 \right) = 0$$

$$y_4 \left(-\frac{9}{40} x_1 + \frac{11}{20} x_2 - 2 \right) = 0$$

$$y_5 \left(-\frac{27}{35} x_1 + \frac{61}{70} x_2 + 2 \right) = 0.$$

But we have been given that $(x_1, x_2) = (80.5, 69.5)$ is the optimal solution to (3) and $(y_1, \ldots, y_5) = (10, 0, 0, 10, 0)$ is an optimal solution to (4). Since $y_1 = 10 > 0$ and $y_4 = 10 > 0$, when the point $(80.5, 69.5)$ is substituted into the constraints of (3), the first and fourth constraints must yield equalities. Conclusion: the policy $D(1) = 1$, $D(2) = 2$ is optimal.

EXAMPLE 2. Consider the Markov decision process of Table 1.2 in Section 1.3, with $\alpha = .75$. (The optimal values determined by value iteration are found in Table 4.3 of the previous section.) With no prior results at hand, we could not be certain that v_3 is nonnegative; but by adding 300 to all the gains, the resulting process would have all nonnegative q_i^k's. The linear programming problem would be to

Minimize

$$x_1 + x_2 + x_3$$

subject to

$$x_1 - .75 \left(\frac{3}{4} x_1 + \frac{1}{6} x_2 + \frac{1}{12} x_3 \right) \geq 393$$

$$x_1 - .75 \left(\frac{5}{6} x_1 + \frac{1}{9} x_2 + \frac{1}{18} x_3 \right) \geq 385$$

$$x_2 - .75 \left(\frac{2}{3} x_2 + \frac{1}{3} x_3 \right) \geq 356$$

$$x_2 - .75 \left(\frac{4}{5} x_2 + \frac{1}{5} x_3 \right) \geq 348$$

(5) $$x_2 - .75 \left(\frac{4}{5} x_1 + \frac{1}{10} x_2 + \frac{1}{10} x_3 \right) \geq 290$$

$$x_3 - .75 \left(x_3 \right) \geq 276$$

$$x_3 - .75 \left(\frac{2}{3} x_1 + \frac{1}{6} x_2 + \frac{1}{6} x_3 \right) \geq 225$$

$$x_3 - .75 \left(x_1 \right) \geq 0$$

$$x_1, x_2, x_3 \geq 0.$$

The solution to this problem is $(x_1, x_2, x_3) = (1477.086, 1376.857, 1297.886)$. Subtracting $1200 = 300/(1 - .75)$ from these values gives the optimal values for the original process, $(v_1, v_2, v_3) = (277.086, 176.857, 97.886)$. The dual problem to (5) has 8 variables corresponding to the 8 inequalities of (5), and the point $(7.72, 0, 0, 0, 2.38, 0, 1.90, 0)$ is an optimal solution. Here $A_1 = 2$, $A_2 = 3$, $A_3 = 3$, and so the stationary policy D, $D(1) = 1$, $D(2) = 3$, $D(3) = 2$, is optimal.

PROBLEM SET 4.4

1. Suppose the constant c is added to all the gains of a finite horizon Markov decision process with terminal values of zero. How are the values of an optimal policy with n steps remaining, the

$v_i(n)$'s, of the original process related to the corresponding values of the modified process?

2. True or false:

a) If, for each i, there is a k_i, $1 \le k_i \le A_i$, such that
$$q_i^{k_i} \ge 0, \text{ then each } v_i \ge 0?$$

b) If, for some fixed i*, all the $q_{i*}^k \ge 0$ for $1 \le k \le A_{i*}$, then $v_{i*} \ge 0$?

3. Use linear programming techniques to determine the optimal values and policies for the Markov decision process of Problem 5 of Section 4.3. In particular relate the optimal solution set of the associated dual linear programming problem to the fact that there are two different optimal stationary policies.

4. As in Problem 3 above, but for the process of Problem 4 of Section 4.3.

5. Use linear programming to determine optimal values and policies for the Markov decision processes of

a) Problem 3 of Section 4.2.
b) Problem 4a of Section 4.2.
c) Problem 4c of Section 4.2.

6. Given a discounted infinite horizon Markov decision process with all gains strictly positive, suppose that $y = (y_1, \ldots, y_A)$, where $A = \Sigma A_i$, is an optimal solution point to the dual of the associated linear programming problem, that is, a solution to the dual of (1). Using only ideas from linear programming, show that

a) The slack of each dual constraint, when evaluated at y, is zero.

b) There is at least one k_1, $1 \le k_1 \le A_1$, such that $y_{k_i} > 0$; and similarly, for each $i = 2, \ldots, N$, there is at least one

$$k_i, \quad \sum_{\lambda=1}^{i-1} A_\lambda + 1 \le k_i \le \sum_{\lambda=1}^{i} A_\lambda ,$$

such that $y_{k_i} > 0$.

c) If y is a basic solution (as defined on p. 81 of [5] or p. 51 of [20]), for each i, there is exactly one k_i,

$$\sum_{\lambda=1}^{i-1} A_\lambda + 1 \le k_i \le \sum_{\lambda=1}^{i} A_\lambda ,$$

such that $y_{k_i} > 0$.

Bibliographic Notes

This monograph contains just an introduction to Markov
decision processes. Other topics that can be studied include nondiscounted infinite horizon processes, optimal
stopping questions, and generalizations, such as processes
with countably infinite state spaces. However these topics
in general are more difficult to develop than the discounted infinite horizon case which we have considered. For example, in the infinite horizon nondiscounted process, one
is faced immediately with the question of defining a hope-
fully finite measure for the value of a policy. One measure that is used is the average expected gain; another is
the limit as the discount factor $\alpha \to 1^-$ of the total expected discounted gain. Also because of the presence of
the discount factor, the important maps T_D and T_α, which we
defined in Section 4.1, are contraction maps, and this sim-
plifies the development in the discounted case considerably.

Books and articles available for further study include
those by Howard [12], Derman [9], Ross [18], Denardo [7],
Veinott [21], Bertsekas [2], and Dreyfus and Law [11].

Howard's 1960 book kindled considerable interest in
the Markov decision process model. In it one will find a
development of the theory for both the discounted and nondiscounted infinite horizon, finite state processes. It is
here that the Policy Iteration Algorithm first appeared in
an infinite horizon Markov decision process setting.

Concerning the other solution techniques D'Epenoux [8] showed that linear programming is applicable in the discounted infinite horizon case, and the bounds introduced in Section 4.3 with the Value Iteration Algorithm are due to McQueen [17]. Discussion of some of the properties of these bounds, and other error bounds, can also be found in the Bertsekas book.

The more recent texts, such as those by Ross and Bertsekas, make use of contraction maps. This approach in a dynamic programming setting was advanced in an early paper of Shapley [19] in game theory, and in papers by Blackwell [4] and Denardo [6].

There are also in the literature various articles developing application areas and real-world models using Markov decision processes. For example, in [13] Klein considers a general inspection, maintenance, and replacement problem for a system or piece of equipment in a state of deterioration, and Howard in his book discusses in considerable detail the replacement problem in terms of the family automobile. DeVries [10] develops a production model as a Markov decision process to determine if and when a small machine tool firm should diversify, and Klein [14] considers the problem of how to compensate for the production of defective items in a manufacturing process. Other application areas of Markov decision processes include hospital admissions scheduling [15], credit control programs [16], and bond refunding decisions [3].

References

1. Bellman, R., _Dynamic Programming_, Princeton University Press, Princeton, N.J., 1957.

2. Bertsekas, D., _Dynamic Programming and Stochastic Control_, Academic Press, New York, 1976.

3. Bierman, H., "The Bond Refunding Decision as a Markov Process," _Management Science_, 12 (1966), B545-B551.

4. Blackwell, D., "Discounted Dynamic Programming," _Ann. Math. Statis._ 36 (1965), 226-235.

5. Dantzig, G., _Linear Programming and Extensions_, Princeton University Press, Princeton, N.J., 1963.

6. Denardo, E., "Contraction Mappings in the Theory Underlying Dynamic Programming," _SIAM Rev._ 9 (1967), 165-177.

7. Denardo, E., "A Markov Decision Problem," in T. Hu and S. Robinson (eds.), _Mathematical Programming_, Academic Press, New York, 1973.

8. D'Epenoux, F., "Sur un Probleme de Production et de Stockage dans la Léatoire," _Rev. Francaise Informat._

Recherche Opérationnelle 14 (1960), 3-16. Translated in *Management Sci.* 10 (1963), 98-108.

9. Derman, C., *Finite State Markovian Decision Processes*, Academic Press, New York, 1970.

10. DeVries, M., "The Dynamic Effects of Planning Horizons on the Selection of Optimal Product Strategies," *Management Science* 10 (1964), 524-544.

11. Dreyfus, S. and A. Law, *The Art and Theory of Dynamic Programming*, Academic Press, New York, 1977.

12. Howard, R., *Dynamic Programming and Markov Processes*, M.I.T. Press, Cambridge, Mass., 1960.

13. Klein, M., "Inspection-Maintenance-Replacement Schedules Under Markovian Deterioration," *Management Science* 9 (1962), 25-32.

14. Klein, M., "Markovian Decision Models for Reject Allowance Problems," *Management Sci.* (1966), 349-358.

15. Kolestar, P., "A Markovian Model for Hospital Admission Scheduling," *Management Sci.* 16 (1970), B384-B396.

16. Liebman, L., "A Markov Decision Model for Selecting Optimal Credit Control Policies," *Management Sci.* 18 (1972), B519-B525.

17. McQueen, J., "A Modified Dynamic Programming Method for Markovian Decision Problems," *J. Math. Anal. Appl.* 14 (1966), 38-43.

18. Ross, S., *Applied Probability Models with Optimization Applications*, Holden-Day, San Francisco, Cal., 1970.

19. Shapley, L., "Stochastic Games," *Proc. Nat. Acad. Sci. US* 39 (1953).

20. Thie, P., *An Introduction to Linear Programming and Game Theory*, John Wiley & Sons, New York, 1979.

21. Veinott, A., "Markov Decision Chains," in G. Dantzig and B. Eves (eds.), *Studies in Optimization*, MAA Studies in Mathematics, Vol. 10, 1974.

Solutions to
Selected Problems

2. (a) States: Yield - heavy (H) or light (L)

$$\begin{array}{c} \\ H \\ L \end{array} \begin{array}{cc} H & L \\ \left[\begin{array}{cc} .7 & .3 \\ .4 & .6 \end{array} \right]. \end{array}$$

(b) States: $ on hand

$$\begin{array}{c} \\ \$0 \\ 1 \\ 2 \\ 3 \\ 4 \\ 5 \\ 6 \end{array} \begin{array}{ccccccc} \$0 & 1 & 2 & 3 & 4 & 5 & 6 \\ \left[\begin{array}{ccccccc} 1 & 0 & 0 & 0 & 0 & 0 & 0 \\ 10/19 & 0 & 9/19 & 0 & 0 & 0 & 0 \\ 10/19 & 0 & 0 & 0 & 9/19 & 0 & 0 \\ 0 & 10/19 & 0 & 0 & 0 & 9/19 & 0 \\ 0 & 0 & 0 & 10/19 & 0 & 9/19 & 0 \\ 0 & 0 & 0 & 0 & 10/19 & 0 & 9/19 \\ 0 & 0 & 0 & 0 & 0 & 0 & 1 \end{array} \right]. \end{array}$$

(c) States: Number of burned-out fixtures the janitor finds
at the end of the month.

$$\begin{array}{c} \\ 0 \\ \\ 1 \\ \\ 2 \end{array} \left[\begin{array}{ccc} 0 & 1 & 2 \\ 81/100 & 18/100 & 1/100 \\ \dfrac{1}{6}\dfrac{81}{100} & \dfrac{1}{6}\dfrac{18}{100}+\dfrac{5}{6}\dfrac{9}{10} & \dfrac{1}{6}\dfrac{1}{100}+\dfrac{5}{6}\dfrac{1}{10} \\ \dfrac{3}{4}\dfrac{81}{100} & \dfrac{3}{4}\dfrac{18}{100} & \dfrac{3}{4}\dfrac{1}{100}+\dfrac{1}{4} \end{array} \right]. \end{array}$$

(d) States: Your point

	1	2	3	4	5	6
1	1/6	1/6	1/6	1/6	1/6	1/6
2	1/2	1/2	0	0	0	0
3	1/3	1/3	1/3	0	0	0
4	1/4	1/4	1/4	1/4	0	0
5	1/5	1/5	1/5	1/5	1/5	0
6	1/6	1/6	1/6	1/6	1/6	1/6

3. Add a new state, 0^*, corresponding to when 0 stoves were on hand at the end of the previous month and a reorder was placed, but the supplier's sources were depleted. In 0^* then, immediate delivery of 4 stoves is guaranteed.

	0^*	0	1	2	3	4
0^*	0	1/5	1/5	1/5	1/5	1/5
0	1/4	3/20	3/20	3/20	3/20	3/20
1	0	4/5	1/5	0	0	0
2	0	3/5	1/5	1/5	0	0
3	0	2/5	1/5	1/5	1/5	0
4	0	1/5	1/5	1/5	1/5	1/5

4. (a) 1

   (b) Once the system enters s_i, it cannot exist.

5. (a) 3/8,   5/8

   (b) 5/16, 11/16

6. (a) For any k, $1 \le k \le N$, the probability of the system moving from s_i through s_k to s_j is $p_{ik} p_{kj}$. Thus the probability of moving from s_i to s_j in two iterations is

$$\sum_{k=1}^{N} p_{ik} p_{kj}.$$

   (b) By induction, the ij^{th} entry of P^n is the probability of the system moving from s_i to s_j in n iterations.

7. (a) $2/3$, $(2/3)^2$, $(2/3)^7$, $(2/3)^{365}$, $1 - \lim_{n \to \infty} (2/3)^n = 1$

   (b) $3/4$, $(3/4)^2$, $(3/4)^7$, $(3/4)^{365}$

   (c) Yes

PROBLEM SET 1.2

1. $q_1 = (1/6)(2 + 3 + 5) - (1/6)(1 + 4 + 6) = -1/6$

   $q_2 = 1/2$, $q_3 = 4/3$, $q_4 = 0$, $q_5 = 1$, $q_6 = -1/6$

2. q_1 = expected gain if last year's yield was heavy

   $= .7(19000) + .3(10000) = 16300$

   $q_2 = 14000$.

PROBLEM SET 1.3

2.

# in stock	# ordered	Exp return	Probabilities 0	1	2	3	4
	4	-80	13/40	3/20	3/10	3/20	3/40
0	2	-12.80	19/25	4/25	2/25	0	0
	0	-80	1	0	0	0	0
1	0	136	9/10	1/10	0	0	0
	2	88	21/50	17/50	4/25	2/25	0
2	0	304	7/10	1/5	1/10	0	0
	2	140.80	11/50	1/5	17/50	4/25	2/25
3	0	376	3/10	2/5	1/5	1/10	0
4	0	400	1/10	1/5	2/5	1/5	1/10

3. (a) $q_0^1 = -100$, $q_0^2 = -56$, $q_0^3 = -80$

   $q_1^1 = 110$, $q_1^2 = 54$

   $q_2^1 = 250$, $q_2^2 = 114$

   $q_3 = 346$, $q_4 = 380$.

(b) The following rows would be added to Table 1.3:

State	Alternative	Exp return	Probabilities 0	1	2	3	4
0	order 1	-52	33/40	7/40	0	0	0
1	order 1	98	5/8	1/5	7/40	0	0
2	order 1	200	17/40	1/5	1/5	7/40	0
3	order 1	254	9/40	1/5	1/5	1/5	7/40

8. States: Value (in $100) of the stock; and one absorbing state, stock sold.

State	Alt	Gain	Probabilities					
			24	23	22	21	20	Sold
24	Sell	2400	0	0	0	0	0	1
23	Sell	2300	0	0	0	0	0	1
	Not Sell	-10	2/5	2/5	1/5	0	0	0
22	Sell	2200	0	0	0	0	0	1
	Not Sell	-10	0	3/10	2/5	1/5	1/10	0
21	Sell	2100	0	0	0	0	0	1
	Not Sell	-10	0	0	1/5	3/5	1/5	0
20	Sell	2000	0	0	0	0	0	1
	Not Sell	-10	0	0	0	2/5	3/5	0
Sold	--		0	0	0	0	0	1

9. States: Level of last year's infestation, if Z, L, or M; HN, last year's level heavy and insecticide not used; and HS, last year's level heavy and insecticide used. Alternatives: spray (S) or not spray (NS).

States	Alt	Costs	Probabilities				
			Z	L	M	HN	HS
Z	NS	100	1/2	1/2	0	0	0
	S	800	1	0	0	0	0
L	NS	300	1/3	1/3	1/3	0	0
	S	900	1/2	1/2	0	0	0
M	NS	800	0	1/3	1/3	1/3	0
	S	1100	1/3	1/3	1/3	0	0
HN	NS	1100	0	0	1/2	1/2	0
	S	1400	1/4	1/4	1/4	0	1/4
HS	NS	1100	0	0	1/2	1/2	0
	S	1600	0	1/3	1/3	0	1/3

12.

States (# reported faulty)	Alt	Costs	Probabilities		
			0	1	2
0	Do nothing	0	.64	.32	.04
	Routine maintenance	10	.81	.18	.01
1	Normal service	25	.405	.49	.105
	Emergency service	60	.81	.18	.01
2	Emergency service	260	.81	.18	.01

PROBLEM SET 2.1

1. (a) $3 \times 2 \times 3 \times 2 \times 3$
   (b) $3 \times (2 \times 3)^{12}$

2. $2 \times (3 \times 2 \times 2)^7$

5. (a) Let $\bar{c} = \underset{k}{\text{Max}} \{c_1, \ldots, c_m\}$.

   Then $c = \sum_{k=1}^{m} p_k c_k \leq \sum_{k=1}^{m} p_k \bar{c} = \bar{c} \sum_{k=1}^{m} p_k = \bar{c}.$

   (b) For the stochastic policy, use alternative k with probability p_k - the expected gain is

   $$\sum_{k=1}^{A_i} p_k (q_i^k + \sum_j p_{ij}^k v_j(0)) \leq \underset{k}{\text{Max}} \{q_i^k + \sum_j p_{ij}^k v_j(0)\} = v_i(1).$$

6. Assume that for some n and all j, $v_j(n) > v_j(n-1)$. Then

   $$v_i(n+1) = \underset{k}{\text{Max}} \{q_i^k + \sum_j p_{ij}^k v_j(n)\}$$

   $$> \underset{k}{\text{Max}} \{q_i^k + \sum_j p_{ij}^k v_j(n-1)\} = v_i(n).$$

   Nonetheless, the claim is FALSE in general. For example, in the following process, $v_1(0) = 10 > 2 = v_1(1)$:

State	Alt	q_i	p_{i1}	p_{i2}	$v_i(0)$
1	1	1	0	1	10
2	1	1	0	1	1

However, if we assume that $v_j(0) = 0$ for all j, then for any i,

$$v_i(1) = \max_k \{q_i^k + \sum_j p_{ij}^k v_j(0)\}$$

$$= \max_k \{q_i^k\} > 0 = v_i(0).$$

Thus, if terminal values are zero, the claim is true; proof by induction.

PROBLEM SET 2.2

1.

n	1	2	3
$v_1(n)$	15	19.6	26.95
$v_2(n)$	8	18.3	23.47
$a_1(n)$	3	1	2
$a_2(2)$	2	1	1

3. (a) $v_1(n+1) = \max_{1 \le k \le 2} \{q_1^k + \sum_{j=1}^{2} p_{1j}^k v_j(n)\}$

$$= \max \{13 + \frac{1}{2} v_1(n) + \frac{1}{2} v_2(n), \ 10 + \frac{4}{5} v_1(n) + \frac{1}{5} v_2(n)\}.$$

Now

$$13 + \frac{1}{2} v_1(n) + \frac{1}{2} v_2(n) = 13 + \frac{1}{2} v_1(n) - \frac{1}{2} v_2(n) + v_2(n)$$

$$= 3 + [10 + \frac{1}{2} (v_1(n) - v_2(n)) + v_2(n)].$$

And

$$10 + \frac{4}{5} v_1(n) + \frac{1}{5} v_2(n) = 10 + \frac{4}{5} v_1(n) - \frac{4}{5} v_2(n) + v_2(n)$$

$$= \frac{3}{10} (v_1(n) - v_2(n)) + [10 + \frac{1}{2} (v_1(n) - v_2(n)) + v_2(n)].$$

Thus

$$v_1(n) - v_2(n) > 10 \Rightarrow \frac{3}{10} (v_1(n) - v_2(n)) > 3 \Rightarrow a_1(n + 1) = 2.$$

(c) Using Parts (a) and (b), if $v_1(n) - v_2(n) > 11.2$,

$$v_1(n + 1) = 10 + \frac{4}{5} v_1(n) + \frac{1}{5} v_2(n)$$

and

$$v_2(n + 1) = -2 + \frac{6}{7} v_1(n) + \frac{1}{7} v_2(n)$$

and so

$$v_1(n + 1) - v_2(n + 1) = 12 - \frac{2}{35} (v_1(n) - v_2(n)).$$

Thus

$$v_1(n) - v_2(n) < 14 \Rightarrow v_1(n + 1) - v_2(n + 1)$$

$$> 12 - \frac{2}{35} (14) = 11.2$$

and

$$v_1(n) - v_2(n) > 11.2 \Rightarrow v_1(n + 1) - v_2(n + 1)$$

$$< 12 - \frac{2}{35} (11.2) < 14.$$

(d) From Table 2.2, $v_1(3) - v_2(3) = 11.3$. Therefore as long as 4 or more weeks remain, alt 2 should be used out of state 1 and alt 3 out of state 2.

6. Always sell at \$2400 or \$2200; sell at \$2300 or \$2000 only if in the last week; and hold at \$2100 for the first week (of the 5 week period), sell thereafter.

8. (a)

n	1	2	3	4
$v_1(n)$	8	$12\frac{1}{4}$	$17\frac{13}{16}$	$22\frac{43}{64}$
$v_2(n)$	1	9	$13\frac{1}{4}$	$18\frac{13}{16}$
$a_1(n)$	1	2	1	2

(b) and (c)

$$v_1(n + 1) = \underset{1 \leq k \leq 2}{\text{Max}} \{8 + \frac{1}{4} v_1(n) + \frac{3}{4} v_2(n),$$

$$6 + \frac{3}{4} v_1(n) + \frac{1}{4} v_2(n)\}$$

and

$$8 + \frac{1}{4} v_1(n) + \frac{3}{4} v_2(n) = 2 + [6 + \frac{1}{4} v_1(n) + \frac{3}{4} v_2(n)]$$

$$6 + \frac{3}{4} v_1(n) + \frac{1}{4} v_2(n) = \frac{1}{2} (v_1(n) - v_2(n))$$

$$+ [6 + \frac{1}{4} v_1(n) + \frac{3}{4} v_2(n)].$$

Thus

$$v_1(n) - v_2(n) < 4 \Rightarrow a_1(n + 1) = 1$$
$$v_2(n) - v_2(n) > 4 \Rightarrow a_1(n + 1) = 2.$$

Now

$$v_2(n + 1) = 1 + v_1(n).$$

Hence

$$v_1(n) - v_2(n) < 4$$

implies

$$v_1(n + 1) - v_2(n + 1) = 8 + \frac{1}{4} v_1(n) + \frac{3}{4} v_2(n) - (1 + v_1(n))$$

$$= 7 - \frac{3}{4} (v_1(n) - v_2(n))$$

$$> 7 - \frac{3}{4} (4) = 4$$

and

$$v_1(n + 1) - v_2(n + 1) > 4$$

implies

$$v_1(n + 1) - v_2(n + 1) = 6 + \frac{3}{4} v_1(n) + \frac{1}{4} v_2(n) - (1 + v_1(n))$$

$$= 5 - \frac{1}{4} (v_1(n) - v_2(n))$$

$$< 5 - \frac{1}{4} (4) = 4.$$

12. The agent should give the assistant Monday and Tuesday off, and call the assistant in on Wednesday only if 4 appointments have been scheduled, on Thursday only if 3 or 4 appointments have been scheduled, on Friday if one or more appointments have been scheduled, and always on Saturday. This policy provides the agent with an expected weekly income of $437.23.

PROBLEM SET 3.1

1.

n	1	2	3
$v_1(n)$	12.5	15.125	18.519
$v_2(n)$	6.5	12.925	15.179
$a_1(n)$	3	2	2
$a_2(n)$	2	1	1

2. (b) $\alpha_0 = 1/2$

6. Use a discount factor $\alpha = 1 - \beta$

PROBLEM SET 3.2

1. (a) 6
   (b) $w_1(D) = 80.5$, $w_2(D) = 69.5$

PROBLEM SET 3.3

1. To maximize total expected gain, the publisher should subsidize a reduction in newsstand price when sales are low and sponsor no promotion when sales are high. This policy yields a total expected gain of 80.5 units if sales this week are high and 69.5 units if sales are now low.

3.

$D(1)$	$D(2)$	$w_1(D)$	$w_2(D)$	
1	1	22.88	21.02	
1	2	20.90	15.74	
2	1	27.38	24.31	← optimal
2	2	23	17	
3	1	18.21	17.61	
3	2	13.45	11.27	

PROBLEM SET 4.1

3. (a) $||x||_a$ is a norm for $N = 1$; otherwise no.
   (b) No
   (c) Yes
   (d) No

5. (a) No
   (b) Yes
   (c) Yes

PROBLEM SET 4.2

4. (a) $D(1) = 1$, $D(2) = 1$ is optimal; and $v_1 = 15.69$, $v_2 = -14.77$.
   (b) $D(1) = 2$, $D(2) = 1$ is optimal; and $v_1 = 17.4$, $v_2 = -13.8$.
   (c) $D(1) = 2$, $D(2) = 2$ is optimal; and $v_1 = 20.14$, $v_2 = -11.66$.

PROBLEM SET 4.3

4. (d) Let D_1, $D_1(1) = 1$, and D_2, $D_2(1) = 2$, denote the two distinct stationary policies. Then Theorem 2, along with the results of Part (c) of this problem imply that both D_1 and D_2 are optimal. Or, using Parts (a) and (b) of the problem,

$$w_1(D_1) = \lim_{n \to \infty} (4 - 3\ (\tfrac{1}{2})^{n-1}) = 4$$

and similarly, $w_1(D_2) = 4$.

7. (a) $\alpha = 1/2$

n	$v_1(n)$	$L_1(n)$	$U_1(n)$	$v_2(n)$	$L_2(n)$	$U_2(n)$
2	16.00	14.00	16.75	-14.25	-16.25	-13.50
3	15.92	15.60	15.83	-14.56	-14.88	-14.65
4	15.80	15.68	15.70	-14.66	-14.78	-14.76
5	15.75	15.69	15.69	-14.72	-14.77	-14.77

9. $5327.36

PROBLEM SET 4.4

4. The optimal solution set to the associated dual linear programming problem is

$$\{\lambda(1,\ 0,\ 3,\ 2,\ 2) + (1 - \lambda)(0,\ 1,\ 2,\ 8/3,\ 7/3)\ |\ 0 \le \lambda \le 1\}.$$

In particular, the point $(1,\ 0,\ 3,\ 2,\ 2)$ corresponds to the optimal policy, D_1, $D_1(1) = 1$; and the point $(0,\ 1,\ 2,\ 8/3,\ 7/3)$ to the optimal policy D_2, $D_2(1) = 2$.

Index